The Railway Age - Preservation Series

FIFTY YEARS DOWN THE LINE

The Story Of Locomotive No.34081
92 Squadron

CONTENTS

This special edition of the **Railway Age Series** ™ is designed and published by

Trans-Pennine Publishing Ltd.

PO Box 10,
Appleby-in-Westmorland, Cumbria,
CA16 6FA
Tel. 017683 51053 Fax. 017683 53558
e-mail trans.pennine@virgin.net
(A Quality Guild registered company)

Reprographics & Printing
Barnabus Design & Repro
Threemilestone, Truro
Cornwall, TR4 9AN
01872 241185

Produced in conjunction and
co-operation with

The Battle Of Britain Locomotive Society

Registered Office
11, Whitehorse Street,
Baldock
Hertfordshire SG7 6PZ

Telephone/FAX 01462 634835

© Trans-Pennine Publishing 1999
Text: Battle Of Britain Locomotive Society
& B C Woods
Photographs: Author's Collection or As Credited

Front Cover: *No.34081 waits to leave Peterborough Nene Valley on 1st May 1999.* **Ian Bowskill**

Rear Cover Top: *On her visit to the Bluebell Railway No.34081 is seen heading towards Horsted Keynes on Sunday 10th October 1998.* **Edward Hurst**

Rear Cover Bottom: *Working an evening train on the Bluebell Railway, No.34081 carries the Night Ferry head-board on 1st November 1998.* **Edward Hurst**

This Page: *No.34081, as yet un-named powers out of Wansford Station in June 1998.*

Title Page: *In this September 1959 view, No.34081 storms up from Ilfracombe towards Mortehoe with a 4-coach Barnstaple train.* **Derek Cross**

FOREWORD

This is all about a rescued and completely restored steam locomotive, and how it was done. A formidable trio, consisting of a steam engine, an illustrious Royal Air Force squadron, and a noted engineer are all crisply introduced. But it is very sad to find at one stage in the story that the engine itself had been abandoned and, no longer wanted, was slowly but surely disintegrating into scrap metal. But then comes a 26-year saga of the rescue and salvage of locomotive No. 34081. Moving it from the dump where it had languished on the South Wales coast to happier quarters, and collecting a small army of dedicated helpers to patiently help with the rescue. It is an inspiring record of innumerable difficulties overcome, financial help cajoled, loyal service encouraged, loyal service maintained, targets set and goals achieved. It also shows the spirit of help given to, and accepted from other restorers. The end became a happy new start, and the re-naming of 92 Squadron with due ceremony included a Spitfire fly-past. And now, fifty years on from its birth, No.34081 is happily back doing a similar sort of work to that my father designed it for all those years ago.

H.A.V. Bulleid

Above: *No.34081 at Exeter Central showing a damaged front cowl, possibly incurred at the coaling point. D.J. Montgomery.*

Below: *H.A.V. Bulleid, at his home in 1997, a very lively 86-year old and son of O.V.S. Bulleid in discussion with the author about this book.*

INTRODUCTION

Many words have been written about preserved steam locomotives, and the format for this book is (by virtue of the task involved) very similar. It is a story of heroic efforts to save an engine from the cutters torch; the struggles against time and money; squabbles within the society set up to save the engine; problems of procuring parts; and finally winning through despite everything. So what's so different about this story?

In some ways very little, but hopefully I have put over the saga of Battle of Britain Class locomotive No 34081 *92 Squadron* in a different light that will make the reading of it both more absorbing and enjoyable.

It is in a different format to similar stories, and we have been liberal with a fascinating range of photographs of the engine (and its sisters), both in BR days and throughout it's restoration. We have included a lot of historical data concerning the whole of the Bulleid Pacific class, as well as the outstanding (some say misguided) viewpoints of the designer himself. In addition a synopsis of his other designs and achievements has also been included. But over and above that this book is a tribute to all those who, against all odds helped get the engine back to running order, in particular **DICK & DORIS MILES**, and to their memory I would dedicate this book.

Barrie C. Woods
Letchworth October 1999

CHAPTER ONE
92 SQUADRON, ROYAL AIR FORCE

It may seem strange to commence a book about a preserved steam locomotive with a chapter giving an abridged history of a famous Royal Air Force squadron. However, my editor considered it the best place to begin the tale, for the locomotive that features in our account was part of the 'Battle of Britain Class' designed for the Southern Railway at the end of World War II. Like many other members of its class it has the honour of bearing the name of an illustrious squadron that took part in the conflict that Sir Winston Churchill described as 'Britain's Darkest Hour.' The squadron concerned was 92 Squadron, which was formed at London Colney on 1st September 1917 as part of the Royal Flying Corps.

Having a crest featuring a cobra snake around two maple leaves, 92 Squadron were soon to live up to their motto *Aut pugna aut morere* (either fight or die). This crest is shown inset on the following page, featuring the plaque from the side of the locomotive. The original personnel were mainly from Canada, hence the Maple leaves, and they initially flew a mixed bag of early bi-planes including Sopwith Pups, SPADs and SE5As.

Above: *RAF 92 SQUADRON on 11th November 1946 at Zeltweg, Austria with one of their famous Spitfires. Note the Alsatian dog sitting on top of the engine cowling. Three of the assembled group attended our re-naming ceremony at Wansford in 1998, 52 years later! Barry Cross.*

Left: *In 1965 following the withdrawal of No.34081 92 Squadron both the nameplates and the crest were presented to the RAF by British Railways. Thus there remained the possibility that, one day, the engine might be re-united with its plates and crests, but this was much more than we could ever dare hope for. However, by 1976, and with our now having established some credibility with the RAF, they decided to present the Society with one set of the nameplates and crests. Tony Fielding (then Chairman) and Dennis Roberts are seen holding the new acquisitions with other locals at the Cheltenham Society of Model Engineers Exhibition at the Bat & Ball Inn, Churchdown. Gloucester Photographic Agency*

with these 'planes they worked up into a scout squadron (what we would now call a fighter squadron), before the embarkation for France in July 1918. In the glorious heat of August that year they went into action in the new offensive above the Somme battlefields. In the intensity of the conflict that followed, 92 Squadron were soon heavily committed and in the months that followed some 37 enemy aircraft were destroyed. They remained on the continent after the armistice, when they became part of the Army of Occupation before the squadron was disbanded at Eil on 7th August 1919.

Shortly after the outbreak of hostilities in 1939, 92 Squadron was reformed from a nucleus of 601 Squadron that was then based at RAF Tangmere. With a re-formation date of 10th October 1939, the squadron was initially equipped with Blenheim 1f fighters. Conversion to the Supermarine Spitfire MkIb fighter soon followed, and the Squadron was declared fully operational on this type by 9th May 1940. This was just in the nick of time, as within a fortnight 92 Squadron were engaged in their first patrol over the French coast.

This sortie saw the destruction of no less than six enemy Bf 109 fighters. Thereafter the pilots were heavily engaged in covering the hasty evacuation of the British Expeditionary Force from Dunkirk, with numerous victories being recorded. After a couple of weeks at Hornchurch, the Squadron was 'rested' and it was moved to RAF Pembrey in South Wales. Here the pilots were engaged in convoy patrols, and they also served as part of the defence force for Wales, the South West and the West Midlands. However, before too long they were back in the thick of the action, after moving to RAF Biggin Hill on 9th September.

The Battle of Britain was arguably the Squadron's finest hour, and it began scoring right from the start. When the battle reached its climax on 15th September, '92' were the first to go into action. Despite this fact, few successes were recorded that day; 'C'est la guere' as the saying goes. Yet by the end of the year, '92' had put all that to rights with the despatch of no less than 127 enemy aircraft, and this was the highest score of any RAF unit in this period.

In 1941 92 Squadron were on the offensive over France. Their conversion to the cannon firing MkVb Spitfire made them the first unit in the RAF to be so equipped, and their offensive role included fighter sweeps and bomber escorts. For the whole of that summer, and well into autumn, the Squadron was engaged over enemy territory on an almost daily basis. In October, and in need of a well-earned rest, 92 Squadron was declared 'non-operational'.

Early in 1942 the Squadron departed for Egypt, where they were equipped with the Spitfire MkVc. On 12th August 1942, 92 Squadron was returned to operational status, and soon found itself covering yet another British military retreat, this time from El-Alamein. Yet the recovery was swift, and in October they were soon on the offensive, this time pursuing Rommel's Afrika Korps across the length and breadth of North Africa. In doing so, life became quite hectic for 92 Squadron as they dashed from airfield to airfield in order to keep up with the rapid British advance. When the spring of 1943 arrived, they were involved in heavy fighting to gain air superiority on the skies over Tunisia. However, the fighting in North Africa soon came to a conclusion, and before long '92' were based at Luqa in Malta.

Their stay here was brief, and despite having only arrived in June, they were kept fully active providing air cover for the Sicilian landings. In July they flew over to Sicily where they were rested in readiness for the Italian landings that would follow at the end of the summer. On 3rd September 1943 the Squadron flew its first standing patrol over the Messina Straits as the landings commenced, but only 11 days later it was flying from Italian soil. By now opposition from the Luftwaffe was very sporadic, and apart from the odd individual battle '92' fought its way up through Italy in what were predominantly 'ground-force support' sorties. It stayed on these duties until the cessation of hostilities in May 1945. In total, the Squadron claimed some 317 enemy aircraft (plus 'probable kills'), and this made it the highest scoring RAF unit in World War II. After a spell on occupation duty in Austria, it was disbanded at Zeltweg on 30th December 1946.

On 1st January 1947 a new 92 Squadron was formed at RAF Acklington from a nucleus of the old 91 Squadron. At this time the RAF were converting to a new generation of fighter aircraft, jet planes. The new '92' was equipped with different variants of the Meteor jet, until it converted to Sabre F4's in February 1954. Thereafter it served with the only RAF Sabre wing based at RAF Linton-on-Ouse, but two years later it was equipped with the new Hawker Hunter. Flying all-blue planes they became the RAF's premier aerobatics display team, and they were given the name 'The Blue Diamonds'. In this role they went on to delight audiences at air shows and displays all over the country, using intricate manoeuvres involving (at times) using up to 18 aircraft.

In 1963 it converted to English Electric Lightning F2s at Leconfield, where '92' remained until December 1965. As our locomotive was withdrawn from service around this time the RAF '92 Squadron' was presented with the nameplates and crests. The squadron then went abroad again, this time to Germany. Retaining their Lightning F2s, they flew from the base at Geilenkirchen until January 1968. At this time the Squadron was equipped with the Lightning F2A, a further move then found the unit at RAF Gutersloh where it remained until it's association with Lightnings finished. In recognition of our restoration project and to cement their association with it, the RAF presented us with one of the nameplates and crests in 1976.

The following year they moved to Wildenrath where they converted to McDonnel Douglas Phantom FGR2 fighters, which were flown in an 'air superiority' role as part of our front line defence and contribution to NATO. In July 1991 the Squadron was disbanded, to be reformed at RAF Chivener in the Autumn of 1992 with BAe Hawk aircraft. They were finally disbanded in September 1994 when it was decided to present us with the other nameplate and crest so that we then had the complete set. The generous nature of this action can be appreciated when one considers that in 1997 one set of plates from a sister engine alone fetched £14,500 at Sheffield Railway Auctions!

Below: *The second set of plates (shown here) were presented to us by the RAF on 12th September 1985, at RAF Wildenrath in Germany. Standing in front of the preserved Hunter in 'Blue Diamond' colours with the plates and a painting of the locomotive are Martin Fox, Ian Bowskill (current Chairman) and Chris Wills. RAF Wildenrath.*

Inset Opposite: *The Squadron crest from the locomotive side.*

Oliver Vaughan Snell Bulleid, pictured inset, was born on the 19th September 1882 in Invarcargill, New Zealand. He was the first child of William and Marianne Bulleid who had emigrated there in 1878. Following his father's death he returned to the UK in 1889 residing with relatives in Accrington. After completion of his formal education some months were to pass as decisions were made regarding his future. A return to New Zealand to study and train as a lawyer under his uncle reached the point of his passage being booked. Fortunately his engineering acumen and the benefits of his remaining in the UK were recognised and he entered the railway industry, where he was destined to become a brilliant mechanical engineer.

Above: *One of Bulleid's stylish light Pacifics from the same batch as* 92 Squadron, *No.34085* 501 Squadron *races through Herne Hill on its way to the Kent coast with the down Golden Arrow. Note its resplendent appointments with the headboard, Golden Arrow side plates, and the twin flags of the Union Jack and the French Tricolour.* British Railways

Right: *A typical example of Southern motive power at Nine Elms depot in 1948. Here we see a pair of 4-6-0s, 443 and 444, which were two of the surviving quintet of the T14 Class. Alongside them is No.34023* Blackmore Vale, *which is one of the West Country 4-6-2s to escape into preservation.* C R Coles

On 21st January 1901 Bulleid began a four-year apprenticeship under Henry A. Ivatt of the Great Northern Railway at Doncaster. Inside just six years he had become assistant to the Locomotive Works Manager. Bulleid was an ambitious character, and he moved to France to take up the position of Chief Draughtsman and Assistant Works Manager with the Westinghouse Company's brake and signal works at Freinville outside Paris. Still anxious to progress, the Exhibitions Branch of the Board of Trade beckoned with the position of Mechanical Engineer. Amongst all this love loomed and he married. However the Westinghouse job petered out, despite the success of the company at the Brussels and Turin Exhibitions in 1910 and 1911 respectively. At this Bulleid came back to England with his young wife and their one-year-old daughter and, on returning to Doncaster, he became assistant to Nigel Gresley until 1915.

Bulleid worked closely with Gresley for many years, but in 1919 he was appointed as Assistant Carriage & Wagon Engineer for the GNR. At the Grouping of 1923, Gresley was made Chief Mechanical Engineer to the newly formed London & North Eastern Railway and Bulleid moved up to become his assistant.

Being of similar age to Gresley, however, Bulleid realised promotion would no doubt be difficult and his active mind required further challenges. These materialised when the post of Chief Mechanical Engineer on the Southern Railway was offered to him in 1937 on the retirement of Richard Edward Lloyd Maunsell. Although rather late in life (he was 55) compared to his contemporaries who gained this elevated position, he attacked the situation with great gusto.

He was soon into everything but his first work appeared in the form of buffet cars on the new electric multiple units, which the Southern were introducing to simplify the intensive working of their system. Many carriage designs, both steam and electric stock, can be attributed to him. Yet despite the move to new forms of traction he carried out considerable work trying to modify and improve existing steam locomotives. In 1941 his first Merchant Navy class 4-6-2 appeared, as we will discuss later. The first member of the class was named *Channel Packet*, and it came complete with a novel numbering system. There is no doubt that Bulleid was influenced by the European system whereby the actual locomotive number was preceded by a code for its wheel arrangement.

Top Left: One of Bulleid's 'Ugly Ducklings' No.33040, the last of the Q1 class, is seen here in particularly good external condition at Nine Elms MPD on 19th May 1963.

Top Left: One of Bulleid's 'Ugly Ducklings' No.33040, the last of the Q1 class, is seen here in particularly good external condition at Nine Elms MPD on 19th May 1963.

Bottom Left: Alongside his first new Merchant Navy Class locomotive, 21C1, Channel Packet, O.V.S. Bulleid stands with Sir Gilbert Szlumper - General Manager, Southern Railway and other dignitaries in March 1941 on the day of the engine's first outing from Eastleigh. National Railway Museum (602-91).

Bottom Right: On a cold 1st January 1948, the first day of the new British Railways, Bulleid No.35017 Belgian Marine heads out of London in a flurry of snow. British Railways

Right: Four months later (on Monday 19th April 1948) driver Flood and fireman Poole from Camden MPD, put the last minute touches to the same engine, after it had been transferred to the London Midland Region as part of the locomotive exchange trials. British Railways

The late-O.S. Nock once commented that this stance was taken by Bulleid, because he held the view that 'one-day British locomotives would be running directly on to the Continent.' As the Southern were the railway that would connect on to any tunnel below the English Channel, this numbering concept was probably more advanced than many commentators have previously given it credit. Anyway, the continental system worked as follows: a Pacific locomotive (with its 4-6-2 wheel arrangement) was classed as 2C1, two axles in front of the drivers, three sets of drivers (A = 1 pair etc.) and one trailing axle. Bulleid juggled these around so the Pacific became 21C, i.e. two front axles, one trailing axle and three drivers. Hence the first Merchant Navy became 21C1 and so on. The same formula was applied to the West Country and Battle of Britain classes, when they were introduced in 1945 and 1946 respectively.

The brilliantly successful 0-6-0 Q1 class, emerged in 1942 but it was rather an 'ugly duckling' and hardly aesthetic looking when compared with the attractive Bulleid Pacifics. In terms of tractive effort these were more powerful than the Pacifics, in fact they were the most powerful locomotives of that wheel arrangement in Britain. The Q1 was undoubtedly a product of the war years, and the starkness of the design was undoubtedly a reflection on the desire to keep down the overall weight of the finished locomotive. The necessity for this can be best appreciated when it is realised that there was an emerging military need for a lightweight, but powerful freight locomotive on the Southern Railway network. We now know that this need was to get quite heavy supplies of war materials down lightly constructed branch lines to small coastal harbours, or the other remote locations where supply dumps were being built up in readiness for the invasion of occupied France.

Above: *Another Bulleid creation that became a white elephant was his 500hp 0-6-0 diesel shunter 11001. With its Paxman Ricardo 12-cyl engine, it had a tractive effort of 33,000lbs, and top speed of over 40 mph. The idea was to have a shunting engine that could also cover trip workings at a reasonable speed. In the event it did neither efficiently. It spent most of its life around Ashford or Norwood Junction but also had trials around Leeds. It was withdrawn in 1959 and cut up at Ashford.*
Below: *Two of Bulleid's prototype electric locomotives emerged from Ashford Works in 1941, the third in 1948 which was visually different and more powerful. All had provision for third rail pick up and over head current collection. The three were numbered 20001 - 3. Seen here, number 20002, ex-works at Eastleigh in May 1959.The late B.J. Swain/Colour Rail (DE 549).*

As many bridges and loading limits precluded the use of traditional heavy locomotives, the Southern's Chief Civil Engineer prescribed an ideal maximum limit of 52 tons. The Q1 Class was the answer, and who cared about its looks in the dark days of war? By way of contrast Bulleid's next major development came in 1945, a year that heralded the arrival of the distinctly stylish West Country class. A year later came the Battle Of Britain class, which is a subject that we will discuss more fully in Chapter Three.

Even so it is worth recording what David Jenkinson states in the book, *Last Years Of The Big Four,* where he writes:

'Because of their 21-ton axle weight, the Merchant Navy's were somewhat route restricted, so this remarkable man set about designing a lightweight version with no more than an 18-ton axle load. By now the elderly nature of many SR types was beginning to become irksome so there can be little doubt as to the logic behind their introduction in 1945; this made sense and since many of the restricted routes lay beyond Exeter, their West Country names were a happy choice. But it was much harder to justify their sheer quantity. By the end of 1947, the Southern had now fewer than 70 in service with 40 more planned...... All told 110, Bulleid light Pacifics emerged to become, amazingly, the most numerous 4-6-2 design ever to be seen in Britain. In the context of the purely Southern requirement, this was patently too many; even in pre-BR days they could often be seen heading but two or three coaches, which was manifestly absurd, the wonder being that it should have happened at all.'

What followed was less successful and his disastrous Leader class in 1946 tempered his earlier success, but even then some component designs were perpetuated long after his departure to Ireland. The bogie mountings for instance, subsequently proved themselves on his main line diesels and furthermore they later did so again on the English Electric Type 4's and Sulzer Peak class diesels. Although many writers criticise the Leader concept, Sean Day-Lewis made the most honest appraisal of these double-ended tank engines in the book *Bulleid The Last Giant Of Steam.*

Despite what is often written elsewhere about Bulleid's desire to perpetuate steam (vis-a-vis the Leader design), it can be clearly shown that he realised that diesel and electric power would be the future of rail transport. In addition to his extensive work on EMU development, he was busy at Ashford with both a modified version of Maunsell's 0-6-0 diesel shunter, and his own unique 500 hp diesel-mechanical shunter No.11001 in 1949. He is also remembered for his Double-Deck electric units. No longer the supremo on locomotive policy after nationalisation in 1948, he was answerable to BR's new Chief Mechanical Engineer Robin A. Riddles. It is often said that this was not the happiest of working relationships, but we will return to this point later.

The year 1950 saw the emergence of the prototype diesel locomotives for the Southern Region numbers 10201-3. In 1954 his prototype main line electric locomotives were out-shopped as numbers 20001-3. But having disagreed publicly with Riddles, Bulleid felt his time with the Southern Region of British Railways was over and he resigned in 1950.

Yet, despite what many commentators have recorded as being a fall-out between Bulleid and his new chief, it has to be said that the two men were in fact long time friends and remained so after 1950. Even though he was now well past retirement age, Bulleid decided to take up the post of Consulting Engineer with the CIE in Ireland. His work there is best remembered for the 'Turf-Burning' locomotive built at Inchicore Works. It's parentage, could be easily traced back to the 'Leader' it's life (for different reasons) followed a similar pattern. Less well known is his substantial work on diesel locomotives, railcars and passenger carriages, where he employed the wide use of welded construction. It was a post that continued on until 1958, his 76th year.

Bulleid was not one to sit back on his laurels and let things happen he had to be there right from the start. He regularly clambered into locomotive cabs on scheduled trains to see how the engines performed, how the drivers performed, and whether they liked their steeds or not. He had his critics, as for example O.S. Nock who wrote:- 'The Bulleid Pacifics were certainly unpredictable and, taken in all, one could sum up their work as brilliant but erratic.' Yet despite Nock's comment, Bulleid's Pacifics have earned an astonishing place in the record books. Almost six decades on from their introduction, they are now the most numerous class of steam locomotives of their type anywhere in the world today. The reason for their preservation is recorded later, but as the late-Ken Hoole wrote 'the fact that no less than 41 Merchant Navy, Battle of Britain and West Country class 4-6-2s will survive to enter the next millennium is truly remarkable.'

The fact that these words have come true twelve years on also demonstrates the dedication of the preservation groups who have rescued Bulleid's creations. Some of Bulleid's other work is discussed in this book to show the astounding brain this man had in developing such a wide range of products which, even if not entirely successful, have certainly left their mark in the history of British railways. His foresight on several occasions paved the way for further development, which was to be successfully exploited and utilised for decades after his departure from the railway scene. His hands-on approach won him the respect and admiration of those who had to work the locomotives that he built. On this point Nock concluded his book *Southern Steam*, 'Bulleid will always be greatly honoured among locomotive men.' At the end of his long career, he moved to Malta where he died on 25th April 1970.

Above: *Bulleid's trio of diesels were fitted with English Electric 16-cylinder engines, the first two (10201-2, introduced in 1951) were rated at 1,600bhp and classed 5P/5F. The 1954 engine (10203) was rated at 6P/6F and had 1,750bhp, generating 50,000lb tractive effort, which compared very favourably against the LMS duo of 10000/10001 that only managed 41,400lbs. This view shows 10203 after a derailment at Norwood Junction in April 1954. Taken from a colour slide supplied by Colour Rail*

Below: *The electrical multiple units built by Bulleid showed a marked improvement on the earlier Maunsell electric units, which employed the bodies of steam-hauled coaching stock. Bulleid's introduction of pressed steel was both innovative and aesthetically pleasing. Southern Railway*

Left: *During World War II, the restaurant and dining car all but disappeared from Britain's railways, as they were deemed a luxury that could not be afforded in those days of heavily laden trains. Many restaurant cars were therefore laid up in sidings and used as office accommodation, etc. Thus, at the end of the war Bulleid had to begin a replacement programme, in which he utilised his pre-war all-steel design to good effect.*

Below: *Leader Class No.36001, having been painted at the Eastleigh Works in June 1949, Bulleid's controversial design is seen on the main line in the late summer. Here the 0-6-6-0 is carrying a party of senior BR officials including Robin Riddles who would eventually make the decision to scrap the project. Notice the ungainly drivers steps, cast 'smokebox?' number plate, cab mounted whistle, Bulleid's electric lights and lamp brackets for headcode discs. British Railways*

Above: *On the resumption peace, boat trains and famous named trains were re-introduced, here 21C119 Bideford (as then unnamed) heads out with what is believed to be the first post-war Golden Arrow working.* Southern Railway official

To understand Bulleid's achievements a little further, we need to gain some perception of his ideology and also work out why he was different from his contemporaries. In fact many Chief Mechanical Engineers would basically go on building what their predecessors had done! They would enlarge a little here; streamline a bit there; tweak up the pressures; improve performance and so on! And, in doing so, they have been very successful!

Bulleid was a man of vision, maybe a little misguided to some, but he was prepared to take the flack if things went wrong. Obviously when designing something you are sure it won't go wrong but if you're prepared to stick your neck out it's a risk you take.

He undoubtedly changed the whole face of Southern motive power, not just steam, but diesel and electric. In addition he was proactive in the field of coaching and wagon stock and in many other areas as well. Obviously he couldn't get everything right, but neither did he get it all wrong.

In this account we must concern ourselves with the powerful Merchant Navy Class that came out in the dark days of World War II, and the 'Go-anywhere ' light Pacifics that followed as peace-time conditions resumed.

It was doubtless an astute move by Bulleid, at the height of World War II, to classify his new 4-6-2 Light Pacific locomotives as 'mixed-traffic' locomotives, thereby circumventing the Government's ruling that no express engines could be produced in the war years. That these engines could do virtually anything is without question so, by dint of persuasion, they were 'mixed-traffic' locomotives! Yet by the same token there can be no doubt that the finished article was none other than a thoroughbred racehorse, and in every way an express engine!

15

Above: *A fine study of No.34081 in her final condition at Nine Elms MPD taken on 29th July 1961, some distortion of the front of the casing is evident.* Jim Oatway.

GENERAL STATISTICS	
Overall Length	67' 4³/₄"
Width Over Cab	9'
Engine weight	86-tons (working)
	77-tons (empty)
Adhesion	56-tons 5 cwt
Tender	42-tons 12cwt (working)
	17- tons 5 cwt (empty)
Total Locomotive	128-tons 12 cwt
Maximum Axle Load	18-tons 15 cwt
HEATING SURFACE	
Evaporative	2,122 sq.ft.
Superheater	545 sq.ft
Total	2,667 sq.ft
CYLINDERS (3)	16⁵/₈" x 24"
Valve Diameter	10"
Valve Travel	6¹/₄ "
Piston Valves	10" dia
WHEELS	
Coupled	6' 2"
Bogie	3' 1"
CAPACITIES	
Coal	5 tons
Water	5,500 galls

Maybe the Merchant Navy class was not of the calibre of the LNER's A4 class or the LMS Duchess class but it certainly was a very close second. Having said that it is important to keep in mind that the West Country and Battle Of Britain classes only had a maximum axle loading of 18ton 15cwt, compared to most other British Pacifics of over 22tons; even Bulleid's Merchant Navy class weighed in at 21tons. Despite this lack of adhesion power the Light Pacifics were still expected to perform like their heavier counterparts! They achieved the tasks set for them, although they did pick up quite a reputation for slipping on starting; but wasn't this a fairly common problem with A4s also? However, they were also prone to slipping at speed, and Nock (commenting on one experience that happened to him at a speed of 75mph) remarked that 'it was not nice at all!'

The West Country and Battle of Britain classes were essentially identical in their statistics. Like all large classes of locomotives variations obviously occurred as time went on, wider cabs to improve forward visibility being a specific example. As with the Merchant Navy class all sorts of alterations were made to the air-smoothed casing and windshields were tried. There were a number of tender alterations and of course the most obvious of all, a major locomotive re-building programme. This came about because of problems caused by the unorthodox front end, failures in traffic (usually the enclosed centre big-end) and their pyrotechnical tendencies. The rebuilt Pacifics were heavier and featured a straightforward front-end layout using three sets of Walschaerts valve gear and a reduction in boiler pressure from 280lbs to 250lbs, with its lowering effect on performance and capability. Yet, the most noticeable feature was the removal of the air-smoothed casing and its replacement with more traditional boiler cladding. In order to appreciate what Bulleid had designed it is interesting to compare the Light Pacifics with other similar top link locomotives.

When comparing the performance tables shown on the opposite page, it is a sobering thought that the Tractive Effort of the ubiquitous Class 08 shunter is 35000lbs! Mind you its top speed leaves a bit to be desired! As can been graphically seen the Light Pacifics were just that, some 15% lighter than their compatriots on other railways! Their pulling (or in particular starting) power doubtless suffered because of this yet once up and running although a little heavy on coal they were amongst the best performers. The huge advantage over all other types shown (with the exception of the largely superfluous 'Clan' Class) was their route availability their axle weight of under 19-tons allowed them to operate over many secondary routes thereby increasing their usefulness considerably.

Loco Class	Wheel Arrangement	Designed by	Year Intro	Wheel Dia	Loco Weight	Tractive Effort	Boiler Pressure
GWR King	4-6-0	Collett	1927	6' 6"	89-tons	40,285	250lbs
LMS Princess	4-6-2	Stanier	1933	6' 6"	104-tons	40,285	250lbs
LNER A4	4-6-2	Gresley	1935	6' 8"	103-tons	35,455	250lbs
LMS Duchess	4-6-2	Stanier	1937	6' 9"	105-tons	40,000	250lbs
SR MN	4-6-2	Bulleid	1941	6' 2"	98-tons	37,500	280lbs
SR BB/WC	4-6-2	Bulleid	1945	6' 2"	86-tons	31,000	280lbs
LNER A1	4-6-2	Thompson	1945	6' 8"	104-tons	37,400	250lbs
NER A2/3	4-6-2	Thompson	1946	6' 2"	101-tons	40,430	250lbs
BR Britannia	4-6-2	Riddles	1951	6' 2"	94-tons	32,150	250lbs
BR Clan	4-6-2	Riddles	1952	6' 2"	87-tons	27,520	225lbs

NOTES:
1) Weights to nearest ton
3) Some weight variations within class (i.e. Rebuilt WC/BB = 90tons)
4) WC/BB & MN boiler pressures subsequently reduced to 250lbs
5) A4 inside cylinder reduced to 17", reduced T.E. to 33616lbs
6) Some variations occur to the above figures

Below: *Taken from a colour slide, this view shows newly rebuilt Merchant Navy Class locos. 35018 and 35020 in pristine condition. Taken in May 1956, the pair pose at Southampton Central prior to their next duties, one of which is the Bournemouth Belle.* The late-S.C.Townroe (Colour Rail BRS 375).

34001: As 21C101 had it's name, crest and scroll painted on - Numerous smoke deflector sizes and shapes were tried on this and various other locomotives to improve forward vision.

34004-6: This trio of engines was fitted with Stanier tenders for the 1948 Locomotive Exchange Trials. The Southern did not possess water troughs thus their own tenders were not equipped with water pick-up apparatus. As longer runs were required during the trials on other Regions this was essential, fitting the Southern engines with Stanier tenders was far cheaper and quicker than modifying their own tenders. They were also fitted with double length smoke deflectors for the same trials.

34004: Fitted with tablet exchange apparatus for working on the Inverness - Perth single line sections., where they had been sent for evaluation.

34005: In 1957, this was the first of the class to be rebuilt, and the work was undertaken at Eastleigh. Sixty members of the class were subsequently dealt with, but despite the photograph in the April 1996 issue of *Steam Railway*, No. 34081 *92 Squadron* was not rebuilt!

34011: This engine was painted in the experimental livery of apple green in 1948, and lined out in grey, red and yellow.

34019/36: As 21C119/36 these engines were to be converted to oil-firing in 1947 at a time when the coal supply situation had led to the decision to convert the sheds at Fratton and Exmouth Junction entirely to oil. The Government-sponsored plan was eventually dropped, due to the lack of the 'foreign exchange' required to purchase the oil. Of the 20 engines designated for conversion, only these two were so treated.

34035/49: Air-smoothed casing modified in further attempt to improve smoke dispersal, but both were returned to standard format

34056: The experimental light green BR livery applied in the period 1948 -1950 as with No.34011.

34057: As 21C157 it was temporarily numbered s21C157 in 1948 to denote the Southern Region of British Railways

34064: As 21C164, this was the first of the class to have a modified cab. Widened from 8'6" to 9'0" with swept-back windows (the narrow cab had been to facilitate working on the reduced loading gauge Hastings line). Light Green livery applied 1948 -1950.

34064: In August 1962 this engine had a Giesl oblong ejector fitted (No.34092 also had one fitted in preservation).

34065: BR experimental apple green livery applied 1948-1950.

34070: The last to be re-painted from Malachite Green to BR Green in 1953, but the first to have her boiler pressure reduced from 280lbs to 250lbs.

34081: This was the last of the class to appear with the front painted numbers.

34082: Smokebox roundel and painted number on front valance discontinued. Cast smokebox numbers implemented.

34086-8: Again painted in the experimental Apple green livery with red, cream and grey lining. However, the lining out was of a different style to No.34011.

34089: This was the last steam locomotive to be repaired at Eastleigh, work that was completed on 3rd October 1966.

Left: *On these two pages we see No.34006 Bude at different stages in her life. This view shows her fitted with a Stanier tender at King's Cross in April 1948 during the locomotive exchange trials. Behind her is the ex-NER dynamometer car, which gained distinction as being the fastest such vehicle ever hauled by steam; it being used in the record-breaking attempt behind Mallard on 3rd July 1938. British Railways*

Right: *Seen at Eastleigh in 1962 fitted with cut down tender and showing the lengthened smoke deflectors she retained to her demise in March 1967, she was one of only three light Pacific's that completed over one million miles. G. Marsh*

34090: Built by British Railways in 1949, but painted in Malachite green with full lining out as a tribute to Sir Eustace Missenden the last General Manager of the Southern Railway. Even the tyres were painted Yellow! The locomotive was also named after him with what was probably the most spectacular nameplate ever fitted to a British steam engine.

34104: May 1961 saw this being the last of the class to be rebuilt.

34109: The first to be fitted with tablet exchange apparatus, for trials on the Somerset & Dorset line. 34040-3 were subsequently fitted with a detachable apparatus and allocated to Bath between 1951 and 1954. There the firemen attached the apparatus as required, and other locomotives were also treated when being sent to this shed for working the heavy summer-time traffic.

34001-80: Originally introduced as 21C101 - 21C180 but later renumbered into the new British Railways system. The members of the class from No.34081 onwards came out in BR days and did not receive the Bulleid numbers.

Entire Class: Tender raves were reduced in height to ease watering, as the bag often caught on the original high sides

Entire Class: Boiler pressure down-rated from 280lbs to 250lbs during 1954 -56 to reduce maintenance.

As with any numerically large class, it was only a matter of time before accidents began to occur, and what follows is the list of known incidents to the class, but there were undoubtedly many others. An example of the type of incident that we have not found details on, are the numerous oil-bath fires to the class, although this book shows pictures of the after-effects on No.34081 and one of her sister engines. What follows is therefore a list of the 'known' incidents.

34020, *Seaton*: On 31st October 1959, this engine ended up in trouble when its driver misread the signals at St. Denys and finished up in the sand-drag at the end of the platform.

34040, *Crewkerne*: On 11th April 1961 the engine was being run out of Waterloo tender-first before being turned. Despite having clear signals for this manoeuvre, the driver had poor visibility to the rear over the tender (a problem when reversing many express locomotives). He therefore failed to see that an incoming EMU had ignored danger signals, and was cutting across his path. The resulting collision claimed the life of the errant EMU driver, and badly damaged *Crewkerne's* tender.

34045, *Ottery St. Mary*: On 2nd September 1961 this locomotive ran through trap points at the end of the down platform at Bournemouth Central station, but without any serious consequences.

Above: *A view of the aftermath at Lewisham following the collision between* Spitfire, *and an EMU in December 1957.*

34066, *Spitfire*; On 6th December 1957 and whilst still fitted with its narrow cab, this engine was involved in the most horrendous accident to occur to any Bulleid Pacific. In fact the collision that took place at St. John's near Lewisham was Britain's third worst-ever railway disaster, and it claimed 89 lives after the driver overran signals and collided with a stationary 8-coach EMU. Both trains were packed to capacity, and if the accident were not bad enough on its own, the severity of the disaster was compounded by the engine smashing into the supports of the Lewisham fly-over, thus bringing the heavy metal bridge down on top of the wreckage.

34084, *253 Squadron*: On 20th February 1960, whilst engineering work was being undertaken to improve the Hither Green Sidings, No.34084 came along the loop line at too great a speed. The points being set to protect the main line it crashed through the buffer stops, after which it rolled down the embankment, causing quite extensive damage.

35020, *Bibby Line*: On 24th April 1953 it sheared its centre driving axles at Crewkerne, shortly after its boiler pressure had been reduced (incidentally, this was the first Merchant Navy to be reduced to 250lbs boiler pressure).

No.	NAME	UNREB/REB	LOCATION
34007	*Wadebridge*	Un-rebuilt WC	Bodmin & Wenford Rly.
34010	*Sidmouth*	Rebuilt WC	Sellinge
34016	*Bodmin*	Rebuilt WC	Mid-Hants Rly.
34023	*Blackmore Vale*	Un-rebuilt WC	Bluebell Railway
34027	*Taw Valley*	Rebuilt WC	Severn Valley
34028	*Eddistone*	Rebuilt WC	Swanage Railway
34039	*Boscastle*	Rebuilt WC	Great Central Rly.
34046	*Braunton*	Rebuilt WC	West Somerset
34051	*Sir Winston Churchill* Un-rebuilt BB		National Railway Museum, York
34053	*Sir Keith Park*	Rebuilt BB	West Somerset Rly
34058	*Sir Frederick Pile*	Rebuilt BB	Avon Valley Rly.
34059	*Sir Archibald Sinclair* Rebuilt BB		Bluebell Railway
34067	*Tangmere*	Un-Rebuilt BB	East Lancs Railway
34070	*Manston*	Un-Rebuilt BB	Sellinge
34072	*257 Squadron*	Un-Rebuilt BB	Swanage Railway
34073	*249 Squadron*	Un-Rebuilt BB	Mid-Hants Rly.
34081	***92 Squadron***	**Un-Rebuilt BB**	**Nene Valley Rly.**
34092	*City Of Wells*	Un-Rebuilt WC	Keighley & Worth Valley Rly.
34101	*Hartland*	Rebuilt WC	North Yorkshire Moors Rly.
34105	*Swanage*	Un-Rebuilt BB	Mid Hants Rly.
35005	*Canadian Pacific*	Rebuilt MN	Mid Hants Rly.
35006	*Peninsular & Orient* S&N Co Rebuilt MN		Glos. & Warwick
35009	*Shaw Savill*	Rebuilt MN	Swindon
35010	*Blue Star*	Rebuilt MN	Colne Valley Rly.
35011	*General Steam Navigation* Rebuilt MN		Binbrook Airfield,
35018	*British India Line*	Rebuilt MN	Mid Hants Rly.
35022	*Holland America Line* Rebuilt MN		Sellinge
35025	*Brocklebank Line*	Rebuilt MN	Great Central Rly.
35027	*Port Line*	Rebuilt MN	Bluebell Railway
35028	*Clan Line*	Rebuilt MN	Stewart's Lane
35029	*Ellerman Lines*	Rebuilt MN	National Railway Museum, York

NB: *Whilst correct at the time of writing, the location situation is continually changing.*

Above: *Preserved Merchant Navy No.35005* Canadian Pacific *seen outside Trans-Pennine Publishing's office near Appleby-in-Westmorland Station.* Alan Earnshaw

Although in her parentage *92 Squadron* was (in every way) a Southern Railway locomotive, her genesis was firmly part of the newly formed British Railways who bestowed upon her the number 34081. It was part of a batch that would carry the numbers 34070-34090, and had the works order No. 3383. She was built at Brighton although most of the Southern works had a hand in her construction; Ashford for instance built both the tender and her main frames. Eastleigh meantime machined the cylinders and supplied the bogies, motion, brake-gear, the steam reverser and cab.

Had production at Brighton progressed as originally planned, the engine would have probably been a fully-fledged Southern locomotive. However delays set in due to shortages of steel and cylinder castings, and this held up production on completion of the previous batch with No.34069 not being completed until November 1947. Therefore, *92 Squadron* was not to appear until the spring of 1948.

Above: *Probably only a few weeks after the engine had emerged from the works, this early view of an as yet un-named No.34081 shown here in excellent external condition, probably some where near Ashford in 1949. The space for nameplate and crest has been left, and it will be noted that the boiler cladding lining stops at the point where these items will be fitted.* Colin Stacey

Right: *Just over a year and a quarter into service and here we see No.34081 (as yet un-named) in her Malachite green livery. (It is interesting to compare this picture with her current livery, which is shown elsewhere in this book). In this instance she is working a Margate - London train on 9th March 1950. Passing West Junction, Ashford, hauling a nine-coach train with a mixed livery of Southern green and the new BR carmine and cream (blood and custard).* Derek Cross

On completion at Brighton Works the engine was paired with a 5,000 gallon tender, but this was later exchanged for one with a 4,500 capacity. Indeed, despite what some readers may think, engines were rarely permanently paired with a single tender. As a locomotive went to the works for overhaul, the tender would usually be separated and the two would part company. For example when 34081 was withdrawn she was coupled to the tender No. 337 from 34028 *Eddystone*. Prior to that allocation, this tender had been originally allocated 34110 66 Squadron. The tender she now carries is therefore younger than the engine, and was built in 1951 to drawing No.W7848 with modifications C40/3451, works order No.10.3486. In 1959 tender modifications were carried out, with the raves being cut down to ease watering and general access.

Locomotive building costs varied a little of course but 34081 originally cost a total of £22,108, of which £3,360 was accounted for by the boiler. Because the engine came under the auspices of BR, the Bulleid numbering system was dropped in favour of the now national standardised numbering system.

In this arrangement ex-Great Western engine numbers were left alone, but it was decided that the other constituent groups would have a prefix to their original number sequence. Therefore ex-LMS engines had the number range 40000 and 50000; ex-LNER locomotives had 60000 added to their numbers, whilst the new standard classes and ex-War Department engines would begin at 70000.

The Southern Railway engines would become part of the 30000 sequence, but Bulleid's idea of CC1 etc would not take kindly to this new system - '36CC1' would have tested any one's patience! So being a 'new BR order' these engines had new BR numbers tagged on to the end of the Southern's list, starting at 34001.

You may be excused from wondering why, having emerged from works several years after the Merchant Navy class, the light Pacifics were not numbered in the 36001 series. Maybe further 'Merchants' were pondered over? Certainly when originally numbered in Bulleid's unique system mentioned above, space was left to accept further members of the class.

The powers behind the new management of BR were strongly influenced by ex LMS personnel, and it was these ones who helped structure the new national motive power policy. One of these new policies dictated that all locomotives would have a cast smokebox door number plate, and cast shed plates on the lower half of the door. The first new Bulleid pacific to be thus fitted was No.34082, making No.34081 the last to come out of the works with it's number painted on the front valance in Southern style lettering. This batch was also the first to carry the new 9' wide 'V' fronted cabs, which gave better forward vision to the crew. They also had the larger tenders fitted.

Feedback from earlier experiences on members of the class already in service led to various modifications being incorporated during the build. These included items on the ash-pan water sprays, and the introduction of flexible main steam pipes because the originals (which were solid), tended to fracture. The front-end frame stretcher layout was modified as were the fire protection plates, and they were given a strengthened smoke box. Longer smoke deflectors were also employed as one of the many attempts to improve driver visibility. The TIA (Traitement Integrale Armand) feed water treatment was fitted as standard, this item being prompted by corrosion problems in the steel fireboxes on the Merchant Navy locomotives.

On emergence from the works she was painted in the livery seen on the cover of this book, with hand-painted 'sunshine' lettering bearing the inscription British Railways on the tender sides. She was originally unnamed, as was the case with most of the class, however whilst some of her sisters enjoyed quite lavish naming ceremonies No.34081 was not to have this honour. As it was not intended to use this locomotive on the Golden Arrow service, the nameplates and crests were fitted close together (where engines were used for the Golden Arrow, the emblem was fixed to the casing between the crest and the nameplate).

In April 1950 No.34081 was repainted at Brighton Works, emerging with a Brunswick Green livery with orange and black lining. The cab side numerals were in Gill Sans, whilst the larger version of the 'lion over the wheel' emblem was applied to the tender sides (with the lion facing forward on both sides). Further repainting works were carried out at Eastleigh in November 1951 and April 1954. During this works visit the boiler pressure was reduced to 250lbs, and the three safety valves re-positioned behind the dome. In 1959 another repaint took place following tender modifications, new lining had to be applied and at this time opportunity was taken to apply the new style BR emblem.

Ramsgate shed (74B) was No.34081's first home, and she stayed there for nearly 10 years during which time she worked the principle express trains including the Continental Boat Train, the Night Ferry, Man of Kent, Thanet Belle and The Golden Arrow although (as stated), she was not built to carry the Golden Arrow emblem between her nameplate and crest.

Above: *Here we have an unusual view of 34081 as it pushes stock through the carriage washing plant at Stewarts' Lane MPD. This picture, taken on 18th May 1956, shows the engine still has the high-sided tender and as yet no front battery box. She was probably on the 471 Ramsgate duty, which involved this operation.* Chris Gammell.

In 1957 she was transferred to Exmouth Junction (72A), which had long had an allocation of West Country class locos. Thereafter she was working around that south western area for a number of years, and certainly hauled the Devon Belle and the Atlantic Coast Express along with other Battle of Britain Class engines. On 1st January 1963 Exmouth Junction MPD was transferred from the Southern to the Western Region, and became a move that saw some 'western features' being added to the engine. A trip back to Eastleigh saw the fitting of AWS equipment which then gave the crews an audible indication as to the state of signals, whilst a speedometer gave them an indication of the actual speeds being achieved.

Below: *In her 10th year No.34081 was allocated to the West Country, and shedded at Exmouth Junction. She was still seen in London with trains from the west, but she also undertook more menial duties. (Our title page picture shows her with a four-coach train from Ilfracombe to Barnstaple in September 1959.) This view sees her in the same area again, with a three-coach local train during that last summer month of the 1950s. Now fitted with reduced raves on the tender and an AWS battery box No.34081 turns her train on to the Ilfracombe branch at Barnstaple Junction. On the right the signalman leans over with the token. Another exile, seen just below the footbridge is an ex-LMS 2-6-2 tank engine. Derek Cross.*

The end of her BR service came at the comparatively young age of 16 years, which was a travesty when one considers that the average life-span of a British steam locomotive was projected at being between 35 and 50 years. However, following the decision to eliminate steam en-mass (rather than strategically and systematically replace it through scheduled withdrawal) some 16,000 engines were hastily consigned to the scrapyard within an eight year period - including many engines that were far newer than the Bulleid Pacifics. As the Southern region steam fleet was to be eliminated before the 1968 national date, No.34081 and her stable mates were progressively consigned to the Eastleigh Works for disposal.

The official withdrawal date for No.34081 was 14th September 1964, but she was sighted arriving at the works with sister engine No.34094 *Mortehoe* on 16th August in a very poor condition and minus her shed plate. Two days later she was recorded as being 'dead' in the Works yard. On the 19th, the Eastleigh Works open day, she was shunted behind the Apprentices Training Building. Presumably out on the premise of being 'Out of sight, out of mind', and thus no source of embarrassment to the hierarchy.

Top Left: *During her stay at Ramsgate No.34081 must have made one of her forays down into the West Country in May 1953. In this picture she is captured at Exmouth Junction MPD, on the road behind Maunsell 2-6-0 N class, No.31836. Of further interest is one of the LMS prototype diesel locomotives, No.10000. Presumably this view was taken during the time when the ex-LMS and ex-SR main line diesels were swopped over for evaluation purposes.* C H S Owen

Centre Left: *Still based at Ramsgate No. 34081 powers through Sidcup on 18th July 1957 with a down empty stock. This working was duty 467 to Ramsgate, but it had been substantially revised due to a serious fire in the Cannon Street signal box. Other pictures of 92 Squadron at work are actively sought by the Battle of Britain Locomotive Society, and we would appeal for your help. If you can help, please contact us at the address shown at the front of the book, or via the publishers. Any loaned material would be copied for archive purposes, and promptly returned. Any such help would be greatly appreciated, as there is still a great deal about our locomotive that we do not know.* John Low.

Bottom Left: *The supply of photographs for this book was particularly difficult, and there were almost as many colour illustrations as black and white pictures. One such colour slide, is now presented as this black and white view of No.34081 as it powers out of Exeter St. David's. The date is May 1961 and here 92 Squadron is on a west of England four coach local, made up of Bulleid set 776.* K.R.Pirt (K.R.Photographics S116)

On Thursday 20th September No.30067, one of the ex-American Army Transportation Corps 0-6-0 tank engines that the Southern Railway had purchased after World War II, dragged *92 Squadron* and four others out of the works yard up to the station. After reversing the convoy into the shed yard, they were left near the coal-stage alongside Campbell Road. She again was shunted around until the 25th when she was espied in the works yard shorn of nameplates and crest. She languished in that poor and gradually deteriorating condition until April 1965, occasionally being shunted as more and more dead engines began building up around the works. As the pressures on the works began to tell, these dead engines were consigned to private scrapyards, many of which were located in South Wales.

Often spring time saw a reprieve for 'stored' engines, and in years past express locomotives might have been repaired and returned to service in readiness for the heavy summer traffic. But there was to be no escape! In April she was coupled up into what seemed to be her last train, along with locomotives Nos.34058, 34067 and 34073, five barrier wagons and a guard's van.

Top Right: *In 1961 she is seen climbing towards Honiton tunnel with a mixed freight. This is the photograph that inspired Malcolm Root to paint his now well known picture of her, albeit with a rake of carriages substituted for the wagons. Yet, at this time, and with a diminishing work load, it was not unusual to see BB/WC class locomotives working very menial trains. We have already shown three/four-coach local passenger trips, and we have views of fellow class members drawing very small freight trains - one comprises just two 6-wheel milk tankers and a brake van. This is surely not the work they were designed for, and is a sad reflection of the bad times that had come upon British Railways in this era! The late-Ivo Peters.*

Centre Right: *On 19th May 1964 the casing has apparently taken some stick, and she looks work weary just five months before withdrawal. As she rests at Exmouth Junction it may have just come off a tender-first or banking duty as there is a rather handsome red hand lamp adorning the nearest electric light. The benefits of electric lamps were obvious but as tender-first or banking work was rare for these engines, red glasses were considered unnecessary. K. C. Fairey*

Bottom Right: *A decidedly grubby 92 Squadron draws into Exeter St. David's in August 1963 with a rake of Southern Region green coaching stock. In an adjacent platform one of the Western Region's diesel-hydraulic Warships is ready to depart for Paddington with a rake of BR Mk.I carriages in the chocolate and cream livery. Colour Rail.*

On their withdrawal date most locomotives would finish their last duty and be returned to their home depot or worked to the nearest large works or major shed. Once there the fire would be dropped, the coal-bunker emptied and water drained from tender and boiler. At the time ours was withdrawn the process of eliminating steam from the Southern Region was well nigh complete and virtually all the depots that were still open were cluttered with dead locomotives. Only a small percentage were actually returned to the BR works for scrapping, because the majority were fully stretched on producing new diesel and electric locomotives to fill the void left by over exuberant individuals glad to rid their area of dirty smelly steam engines.

Consequently it was necessary to put the scrapping of steam out to tender to private yards, and this is a story well told in a book written by my editor (Professor Alan Earnshaw) and entitled *Steam For Scrap - The Complete Story*. Briefly told, however, the story of the demise of British steam has an important part in our story of No.34081.

By now it was a redundant asset that BR wanted to dispose of as quickly as possible. As with all tendering the best price won the day and convoys of steam engines would be dragged all over the place to the appropriate yard. Regional boundaries that had stood the test of time were rapidly demolished as an NER Q6 found its way to Friswell's of Chesterfield. A Schools class locomotive, No.30935 ended up at Cohen's in Kettering, whilst King's of Norwich were host to GWR Castles and Counties. Dai Woodham's Yard, at Barry, as is well known, boasted engines from all four major regions including No.34081.

However, Woodham's had also gained a contract to cut up thousands of redundant coal wagons (which were actually more profitable than locomotives), so No.34081 together with some 200 odd other (fortunate?) locomotives were left relatively intact. The tender was a different matter however, and that fitted to *92 Squadron* was sold to the Britton Ferry Steel Works. Stripped of its tank, the wheels and frames were thereafter used as a billet wagon.

Left: *A sorry sight indeed at Eastleigh on 7th November 1964 as No.34081 now shorn of number and nameplates, but otherwise intact awaits her fate as the private scrap yards around the country put in their bids for engines. It was this lottery process that decided those that would live and those that would die. The lucky ones of course went to Woodham's at Barry, but other yards would ruthlessly despatch new arrivals into wagon loads of scrap. Of the BB/WC class just nine were cut up at the Eastleigh Works (Nos.34011, 34035, 34043, 34049, 34055, 34068, 34069, 34074 and 34110). Cashmores at Newport cut up 38, Buttigeig's in the same town despatched 18. Along the South Wales coast Birds cut up nine at their Llanelly yard, six at Bridgend and three at Morriston. One more went at Woodfield's at the Town Dock, Newport, and four died at the yard of Wood's in Queenborough, Kent. C. Stacey.*

Below: *April 1965 saw No.34081 and three other dead Bulleid locos being towed to Woodham's by sister engine No.34006 Bude. The irony was that of the five locomotives in the convoy, the only one still in steam (which was also the only one with any distinctions, having been involved in the Locomotive exchanges of 1948) was the only one of the five to be cut up. The reason behind this was that No. 34006 Bude was one of the 38 Light Pacifics that ended up in the Cashmore's scrapyard at Newport where little time was lost in demolishing any new arrival. Although some may think that Woodham's saved all the engines that went there, they did not (for they were in the business of re-cycling scrap). In fact two members of the class met their end at Barry, these being No.34045 Ottery St, Mary and No.34094 Mortehoe which, all considered were really unlucky engines. The late-Ivo Peters*

To those who journeyed there this roof top view of Barry scrap yard was a must. So many of us failed to realise the rapidity with which steam was disappearing from the railways. But who would ever have imagined then that nearly every locomotive in this view would be saved? Steve Worrall.

The engines and tenders left at Barry sat in long rows gathering dust, rust, and moss as the corrosive sea air aided the gradual degradation of components. Yet it was not only the elements that caused problems, for the biggest threat was the more obtrusive hands of bounty hunters who systematically robbed the locomotives of all saleable items particularly copper piping. The problem or advantage of Woodham's site at Barry Island, according to your aspirations, when compared to most other scrap yards was that it was an open unfenced area and access was virtually unimpeded. During this period the author made numerous trips to Barry Yard, one on the 17th March 1968, five months prior to the end of steam. The following Southern Region locomotives were there:-

Barry (A Dump)

30830	31618	34007	34010	34016	34027	34028
34039	34046	34053	34058	34059	34067	34070
34072	34073	34081	34092	34101	34105	35005
35006	35009	35010	35011	35018	35022	35025
35027	35029					

Barry (B Dump)

30499	30506	30825	30828	30841	30847	31625
31638	31806	31874				

In all a total of 40 engines at that time, all of them (plus a later arrival No.30541 were all preserved), making a total of 41 engines.

Early Days: *Colour images of No.34081 in the first ten years of its life are very rare indeed, and the Battle Of Britain Locomotive Society would love to hear from anyone who can help.*

Above: *How the Battle of Britain class locomotives looked when they were new. In lieu of No.34081, we show s21C157, Biggin Hill, on the down 'Thanet Belle' near Henley in July 1948. She already has British Railways on the tender and a splendid rake of slab-sided Pullmans in tow. Friends of the National Railway Museum/Colour Rail (BRS 1063).*

Right: *A wonderful composition, and one of our earliest colour pictures of No.34081 taken in 1959 departing from Southampton Ocean Terminal. Bus fans will no doubt drool over the Duple-bodied Bedford SB coach. Colour Rail.*

In 1973 an initial approach was made to Dai Woodham by a pioneering group to request permission to inspect the Bulleid Pacifics to select one for possible preservation. This pioneering group comprised of Tony Fielding, Dennis Roberts, and Andy Hiles. Although looking thoroughly weather beaten and with many parts missing, including major items like the coupling rods and tender, No.34081 had a sound boiler (No.1288) and frames. After more discussions and another visit, a working party was despatched to Barry to commence remedial work in readiness for the loco to be moved and to try to halt the ingress of corrosive sea air. At this time the momentous decision to purchase was made and a cheque for £3,850 was handed over to Mr. Woodham for that purpose.

The embryonic Society joined the ARPS (Association of Preservation Societies) and 300 Shares at £1 each were issued in the locomotive. Serious acts of fund-raising were now in full swing to raise the cash to pay for the loan covering the purchase and secondly to pay for the inevitable move to wherever. One highlight of the year was a charter train, the 'Torbay Express', which was run from Gloucester to Paignton.

It was quite a successful venture, and carried 300 passengers. Our links with the RAF were being forged early on with Air Vice Marshall R. W. G. Freer CBE, MBIM, RAF of AOC 11 Group, becoming our Patron whilst Wing Commander C. W. Bruce, of OC 92 Squadron, took on the role of President. Meanwhile Fl. Lt. J. K. Fletcher (ret'd) presented the Society with a painting entitled 'Return to base'. By now Tony Fielding, the original Chairman, had developed an enthusiastic group in the Gloucester area, whilst Tony Missen had done the same in Wiltshire. John Dennison had formed a further group in Nottingham, and other members were spread far and wide across the country.

However, a shift of emphasis with regard to the location of the Society began to materialise around this time with the involvement of a certain Michael Watts who lived in Letchworth. In due course he became head of what was temporarily known by the grandiose title of 'London Area Group', and in an attempt to increase local support he placed an advert in the local paper. In turn this brought quite a few new members into the fold, including yours truly.

Left: *This wretched-looking engine at Barry is what our intrepid pioneers had let themselves in for! By the time the purchase was arranged her tender had gone for use at a Welsh steel works, so another was procured from the dump at Barry Island. The engine meanwhile had been robbed of all non-ferrous parts, coupling rods and many other items, what was left was bent, battered and rusty. A long restoration project lay ahead, but it is worth comparing this picture with that on the front cover to appreciate what has now been achieved by a dedicated band of volunteers. Steve Worrall.*

Below: *The piece of paper that started it all, a receipt for the sale of No.34081 signed by Dai Woodham on the 27th September 1973. Even though this meant that the engine was finally ours, it was to be another three years before we were able to move her away however.*

Our AGM's were held in the RAF Museum at Hendon, which if not central to the group was at least appropriate on other grounds. The Museum had been most helpful with the production of a first day cover featuring a 92 Squadron Spad and a Lightning. Committee meetings were held at RAF Lyneham in Wiltshire which apart from Tony Missen (the lucky one!) meant horrendous journeys for most of the others coming from the West Country, South Wales, Gloucester and of course Letchworth. As one would expect our arrival at the RAF base was subject to a concerted identification check of both car and occupants, following which we then drove a good way around the airfield passing close to lines of Hercules heavy transports! I doubt that in today's more terrorist dominated environment that would ever happen! The meetings were held in a World War II Nissen hut. A long return journey followed, and arrival back home in Letchworth at 2.00am was not unusual!

Working parties were arranged to go to Barry to prepare the loco for its eventual move. One such party set out on the 24th November to jack up the cab and insert mahogany blocks between the axle boxes and trailing axle to serve as temporary bearings. Similar work was done to the tender; a hard enough job in itself, but one that was made all the more difficult when one realises that No.34081 had been parted from its tender on arrival at Barry. The substitute tender we had purchased was about 100 yards away from where we could park the van whilst the loco was 200 yards away! This exercise was no fun at all, especially when you are lugging a 40-ton jack across rough ground!

Enthusiasm was further strained when it was discovered that since the last visit someone had made off with two sets of tender springs and the rest were unbolted ready for removal! This had been done despite the fact that large signs on the tender proclaimed our ownership, so acting on the premise of being better safe than sorry, the second day of the visit was spent removing the remaining springs for safe keeping.

Following the purchase more work was carried out in 1974. The pony truck bearings were protected, whilst the axle keeps were jacked and greased in order to prevent damage in that area. Canvas covers were placed over what was left of the casing. Dia Woodham arranged for the local 08 shunter to move No.34081 out of the storage line and on to the departure siding, but this was never going to be an easy task as the locomotives were often completely seized after 10 -15 years in the yard. This problem is testified to by the well-known case of 'King' Class No.6023, which had the lower half of its driving wheels cut away.

Fortunately No.34081 was successfully shunted and placed with B1 61264 and a couple of others in readiness for collection. The Society also became registered for VAT, which brought a very welcome £350 back into the coffers from the purchase price!

West Country Days: *Here we present views of two Bulleid Light Pacifics (now preserved) that will come as a complete surprise to Bulleid enthusiasts, as we show two wonderful colour shots that have never been seen in print before.*

Above: *In July 1961, No.34081 is working a down Atlantic Coast Express relief, and calls at Okehampton station with a rake of SR green coaches on a misty Saturday afternoon. The luggage trolley on the opposite platform adds to the period scene.*

Left: *Exactly twelve days later, but this time we see No.34073 249 Squadron as it simmers quietly in the sunshine at Exeter Central Station, as it waits to leave with a train for Bude.*

Both pictures are from the Trans-Pennine Archive, which comprises some 17,000 rare unpublished colour images.

Hot Days: *It is well known that the Bulleid locomotives were prone to oil-bath fires, but just how serious these fires could be may not be widely appreciated. Therefore the two images on this page act as timely reminders.*

Above: *Obviously, from the repainted casing panels, No.34081 appears to have suffered an oil bath fire. When spotted at Exeter Central with the 15.00 from Waterloo on 20th May 1963, the smoke deflectors carry the rails for fitting of the 'Devon Belle' headboards (or should they be called sideboards?). R. Leitch/Colour Rail.*

Right: *Evidence of just how bad the fires could be are seen on 34049 Anti-Aircraft Command at Eastleigh in 1954, following what appears to have been a lagging fire. The late B. J. Swain/Colour Rail (BRS 339).*

By 1975, apart from the mainframes, the boiler was becoming the main worry. As with all steam preservation schemes acquiring engines from a scrapyard, no protection was offered to a locomotive once it had been despatched there. So a professional in the field, Mr. H. Smart, was engaged to check ours out. Thankfully he pronounced it, and the whole engine, as sound. Comments were added that there were a lot of parts missing, but then that applied to all such engines. So with an authoritative opinion to boost morale, fund-raising and preparatory work continued unabated throughout the year. We slowly paid back the loan on the purchase price and raised additional cash for the transportation. All of this caused our then Treasurer, Stewart Kevill-Davies, continual headaches. These eventually got so bad he had to migrate to Australia in order to get away!

The next obvious question was where was it to go? We all wanted it to be within a reasonable distance from our homes, which meant that three miles north-west of Stow-on-the-Wold would have been, geographically, the most ideal place! Much heated debate ensued, but in fairness the debate concerned the facilities that a particular railway could offer, rather than personal whims. Societies considered included the Bristol Suburban Railway, the Dowty Railway, Aschurch, RAF Quedgeley, the Dart Valley Railway (a strong contender), the Nene Valley Railway, the Main Line Steam Trust, the Bluebell Railway, and the Dean Forest Railway (another strong contender). Some of these were rejected almost immediately, but questions were raised with others. Such as, whether they would like '92'? Could covered accommodation be guaranteed? What were the railway's future plans?

The Dart Valley Railway very nearly became the home for No.34081 but lack of any firm moves by them to secure an agreement with us persuaded us otherwise. Eventually the Nene Valley Railway at Peterborough was chosen. Detailed discussions then took place to arrange for the procedure of getting '92' physically on to the line, and for accommodation at Wansford.

In 1976 transport companies were approached to ascertain costs for removal of the loco and tender. Two low-loaders would be required for an exercise that, with loading, would take a couple of days to complete. Wrekin Roadways provided the transport in the shape of two Foden heavy haulage tractors. The loco and tender were winched on to their respective low-loaders in the conventional manner although at one point it was interesting to see one of the tractors dragging '92' along by itself to bring it into position for loading. The journey was made over-night, and the convoy arrived at Peterborough as the sun rose on the morning of 6th November 1976. Two hastily prepared notices adorned the engine (which was loaded on rearwards) one hanging from the cab, facing forwards proclaimed:

34081 92 SQUADRON,
BARRY TO PETERBOROUGH
NENE VALLEY RAILWAY.
Whilst one hanging from the smokebox at the rear stated:
I'M GOING TO PETERBOROUGH
NENE VALLEY RAILWAY.

Thus adorned No.34081 was the 86th locomotive to leave Barry Island. Off-loading at Wansford was out of the question due to lack of space and facilities, but at the other end of the line the British Sugar Corporation sidings at Peterborough were made available for our use (as had been done with earlier arrivals). Once back on terra-firma No.34081 was marshalled into a short train with a couple of box vans for the very slow trip to Wansford Yard.

The train was hauled by Malcolm Heugh's 0-4-0ST Barclay, Works No. 2248, built in 1948, the same year as it's load! Interestingly the Barclay had previously worked at other sugar factories at Wissington and Kidderminster, but Malcolm then purchased it for preservation and numbered it 90432. The reason behind this was that his father had been a driver at Boston MPD and the ex-WD 2-8-0 No.90432 was the last engine he drove. It was a tense time as apart from the odd shunt over a few yards this was the first time the engine had moved any appreciable distance in 11 years, but No.34081 arrived safely without too much falling off or seizing up!

Left: *The scene at the Barry Dump in September 1976, some two months before No.34081 was to depart. Here she is seen sheeted over and on the departure line with a GWR locomotive bound for the Dart Valley Railway.*

Above: *The scene at Woodham's Yard on 7th November 1976 with the locomotive ready for removal to the Nene Valley Railway at Peterborough. Although the engine looks in an awful condition, this belies the amount of work that had been expended on her at Barry in readiness for her transfer to Cambridgeshire. Although most of this work was of a 'needs must' nature, valuable items (that might otherwise have been robbed) were removed for safe-keeping.*

Right: *After a long overnight journey No.34081 arrived at the British Sugar Corporation factory in Peterborough, shortly followed by her tender on another low loader on November 6th 1976. A month later she had been coupled to the tender again, and once united it was time for her first move on the Nene Valley Railway. On 7th December, she was very carefully dragged to Wansford by Malcolm Heugh's 0-4-0ST Barclay shunter which, incidentally, was built in the same year as No.34081. Ian Bowskill.*

37

Tragic Days: *The early days at Barry are not that well recorded in colour images, so these two pictures from the Trans-Pennine archive are again an interesting addition.*

Above: *In May 1965 No.34070 (formerly 21C170)* Manston *sits in a dilapidated condition at Barry amongst a clutch of GWR tank engines.* Manston *arrived at Woodham's yard some four months before* 92 Squadron, *but their condition at this time would have looked very much the same.*

Left: *Some five years later, our photographer returned to Barry, and the scene on a wet August day - twelve months on from the end of steam. Beyond the line of scrap coal trucks, a batch of Bulleid Pacifics are also seen, and this line up probably includes* 92 Squadron.
Both pictures Trans-Pennine Archive.

Hopeful Days: *As the preservation movement finally got its act together, the dump of locomotives (deteriorating in condition all the time) saw its members gradually being ear-marked for salvation.*

Above: *This view is fairly representative of what the engines looked like by the early 1970s, this view was captured by Alan Earnshaw on one of his trips in 1972. Even now it is hard to imagine that over a quarter of a century has elapsed since this view was taken.* Trans-Pennine Archive

Right: *How sweet it all seemed as No.34081 rolled into Peterborough on 6th November 1976. This was true in more ways than one, for at the end of a long overnight journey she arrives at the British Sugar Corporation factory where unloading would take place.* Ian Bowskill.

Above: *Maybe not the pioneering group and certainly not the most handsome lot! But the 1977 core of the Society, many of whom are still active today. It includes Roy Tanner, Graham Hall, Richard Perkins, Barrie Woods, Ian Bowskill, Martin Fox, Glyn Sims, Derek Butler, Alan Whenman, John Arthur, Kevin Wilkins, Dave and Sally Capon.*

In 1977 the work really began, stripping and sorting parts under the auspices of Alan Whenman our Engineering Officer. Whilst we were busy on the loco the relatively unknown Malcolm Root was busy on his canvas and produced an excellent painting of *92 Squadron* from a photograph by Ivo Peters. Prints were taken off and sold for £4 each. Unfortunately the original was subsequently lost to us in one of the fairly rare disputes within the Society whereupon an officer decided to retain it to recompense his alleged loss. Hopefully, some day reason will prevail and we will have it back where it belongs. Around this time 92(F) Squadron appointed a new commanding officer, Wing Commander E. Durham and he duly accepted the Presidency of the Society. A model railway exhibition was held in Letchworth on 30th October which included a wide range of exhibits including some dustcarts measuring nearly two feet long made entirely of matchsticks! £170 was raised for the funds.

By 1977 our Sponsorship Officer (the late-Graham Hall), was very active during these early years as although sales were bringing a healthy return, engineering costs would rapidly out strip the income. Sponsorship therefore was important and many companies were wonderful in being persuaded to part with materials, goods and machinery to help us along.

The list included hot water washers, needle guns, portable diesel compressor, air tools, slings and tackle for boiler lift, shot blasting equipment, mobile welding unit, blasting grit (by the ton!) and paint, all of which assisted in the initial preparation work and saved the Society thousands of pounds.

During the dismantling and initial renovation work most components removed had to be cleaned down in some form or another. This work revealed an interesting side issue that has occurred on the majority of preserved engines. It is the realisation that what you have purchased is not quite what it purports to be. Number 34081 was no exception, as each part was inspected the policy of railway companies to stamp all components with the receiving locomotive's number was revealed.

The policy as far as No.34081 was concerned was a little perplexing, as parts from the other locomotives were found. Whilst it is easy to put the facts down to human error when re-assembling, we could find no explanation as to why all the numbers are from early members of the class (including Nos.21C102, 21C108, 21C114, 34018, 34022, 34024, 34032, 34037, 34041). We wondered, could it be that these particular engines were in works for attention whilst ours was being erected? But then we knew ours was built at Brighton and repairs were mainly carried out at Eastleigh.

However, it transpires that in the latter part of the 1950s a policy of remove and replace was instituted at many of the works. This meant that certain components were taken off an engine when it went in to the works for repair, and replaced with spares. The removed items would then be refurbished and placed on to the next engine in for repair, and so on. Thus the numbers stamped on locomotive components would often be from those with lower numbers. The same practice also happened at the main sheds, where components were occasionally swapped from 'out of service' locomotives in order to keep other engines moving.

On-site accommodation was considered at this stage as many volunteers had to travel quite some distance to Wansford, this all took time and petrol money. Eventually a 44ft x 10ft mobile home was acquired from Clifton, Beds, after a bit of haggling for just £80. Stewart Kevill-Davies then did some smart work on a wholesale stationary company, Porter Brothers from Wimbledon, who kindly paid for it. The size of it precluded towing on the road, so Ivel Transport in Biggleswade were persuaded to move it on one of their specially fitted out low-loaders, again free of charge. It was taken to Ashwell, a village near Royston in Hertfordshire, a distance of about 15 miles. At Ashwell we had approached Chamberlain Crane Hire who had their base in the old station yard. They allowed us to park the 'van there whilst it was renovated.

As I lived in Ashwell it seemed logical for me to take on the project. To say it needed a little work doing on it is to slightly under estimate the job. If you stood outside the van at one end and looked in the window you could look right through the gutted interior and out the other end! Renovation on it progressed, various types of materials were sought (scrounged) and furniture was added.

Then we had a break-in, and the recently acquired three-piece suite was stolen. It was time to reconsider our location. Fortunately a new member, Dick Miles, took up the cudgels, for he had a farm near Baldock just a few miles away and kindly said that he had space for the van. Ivel Transport again assisted and transported it to the farm. On route one of the side panels began blowing about a bit in the wind, remembering the thing was 10ft wide this caused us some consternation, but we arrived safely. However, backing the unit into the farm was a lengthy process, which entailed completely blocking the main A505 road for some time. During this operation a police patrol car turned up, and we dreaded the worst. However, it was refreshing to find that the officers concerned were exceedingly helpful in assisting in what could have been a rather difficult situation.

Firmly ensconced behind the pig-shed work re-commenced, with Dick and myself doing the bulk of it in the evenings and at weekends. It was a good time and we progressed well over the next couple of years, but during that time I learnt more about pigs than steam engines! With the engine at Peterborough interest shifted quite naturally, waning in the West where it had all started but taking on a new impetus in the South East and Midlands. Mike Watts was appointed Chairman at the Hendon AGM, and with this shift Committee meetings were eventually held in his home at Letchworth. Apex Engineering of Letchworth heard of our project and, following a discussion they offered to rebuild the cab for us. Having already been shipped down to Dick's farm from Wansford it was but a short distance to their premises.

On the fund-raising side many sales stalls were held and these raised a considerable amount. Another model railway exhibition was organised in Letchworth, which really went with a bang, especially when one of the sales stands collapsed! Even so some £380 was raised that day. This year saw the purchase of a set of coupling rods at £2,000 and a pair of connecting rods at £650 each from Clark's Crank & Forge Co. Ltd. A buffet dance was held in Letchworth and sales were doing well at traction engine rallies, air-shows, raffles and the like. Engineering were still stripping parts but, more importantly, they were beginning to renovate them as well.

By 1979 the tender came under scrutiny with both frames and the tank being grit-blasted. Scrutiny also came from the group restoring the unique locomotive No.71000 *Duke of Gloucester*, having been suitably impressed by our efforts with the grit blaster they went ahead and purchased their own machine. (Beat us to it also by a number of years didn't they!)

Dick's Farm, apart from the caravan, was now housing many parts off the engine. It was commonly thought that at this period of time there was more of the engine at Baldock than Wansford! The cab panels had been kindly made and painted by Apex Engineering of Letchworth but not fitted at this time.

Exciting Times: As the long slog came to an end, we began to see 1998, 34081's Golden Jubilee year, as one of the most exciting times in her history. After 30 years, she slowly came back to life.

Above: Many of us wondered if this sight would ever happen, No.34081 engulfed in steam! Yet by 23rd March 1998, she was running bedding-in trials up and down the Nene Valley line.Our big dilemma was when to run the engine and when to paint it! Compromise ensued and at this stage, the loco is in undercoat but the tender has been finished and sign-written.

Left: It is said that every picture tells a story, and every picture is worth a thousand words, but this (for us at least) is a picture that simply says it all. Ian Bowskill

The wheels and journals were ready for collection from Swindon Works where some excellent work had been done under the watchful eye of John Sandford. We also had the bearings re-metalled, new safety valves, cylinder cocks, and injectors. New drawgear, valve gear chain wheels and chains were also produced. At Wansford, clearance work enabled us to erect a shed for storage of heavy equipment, a problem that had been nagging at us for a long time. Debate on the livery was quite heated with a variety of strong views held. Some differences of opinion on this and some other matters led to the resignation of Chairman Mike Watts, so Dick Miles stepped into the breach. It was finally agreed to put the livery question to the members and a vote with three basic options ensued:

1) SOUTHERN LIVERY: Malachite green with yellow stripes, 'Southern' lettering on the tender sides, a Southern brass ring on the smokebox and the Number 21C181. This was not authentic as nationalisation intervened prior to the emergence of the engine from the works as No.34081.

2) LIVERY AS BUILT: Malachite green with yellow stripes British Railways lettering on the tender sides in white, no brass ring and No.34081 painted in SR style on the front valance and BR style on the cabsides. Nameplates were not fitted whilst the loco enjoyed this livery but we were to add these.

3) BRUNSWICK GREEN LIVERY: This was the standard BR express locomotive livery with orange and black lining, a cast smokebox door number plate, lion & wheel motif on the tender sides and 34081 on the cabsides

Members were advised of this in newsletter No. 18 that September and given until the end of October to decide. The following newsletter printed the results:

1. SOUTHERN LIVERY	12 votes
2. AS BUILT	42 votes
3. B.R. BRUNSWICK GREEN	39 votes

The total of 93 votes represented about half the membership, presumably the other half didn't mind or care what colour it was to be! On the subject of the livery, some other more prosaic suggestions were received from the lunatic fringe:- Luminous green smoke-box door with purple metal-flake casing and a supply of pink buckets for the visitors to be sick in! Another suggested 'Puce' with lilac lining and frilly bits around the edges! Yet another requested that we scrap the engine and with the funds build a new Gresley P2. Maybe we do get too serious about our hobby sometimes!

In addition to the 'paint debate' engineering work progressed, the tender class 'F' brake cylinders were completely overhauled and the piston rods replaced by new ones. The tender body was almost ready to mount back on to the chassis, but we were still awaiting the return of the wheels from Swindon. Following an unofficial inspection by a BR expert, it was decided that the existing tender axle boxes could be re-used. He recommended that the vacuum brake cross shaft be equipped with a proper grease lubricator rather than rely on the driver's oil can, and also suggested some pipe work could be adjusted to improve maintenance.

Above: *Work on the tender (formerly from No.34028 Eddystone) was undertaken to hold its condition whilst the main work on the locomotive progressed. Here split from wheels and frame the 'as bought' condition of the tank can be appreciated. Richard Perkins.*

Below: *A few months later after a good deal of grit blasting, scraping, filling and painting the tender had taken on a completely new appearance. There is still a long way to go, but as the empty paint cans testify, an awful lot of really good work had been achieved by the time the picture was taken. Richard Perkins.*

Above: *With the tender undergoing restoration down the yard a pause for reflection by two stalwarts, enjoying a glorious summer day in 1977, in between unbolting the cab. The ubiquitous Mk.III Ford Cortina adds to the period scene - was it really over 20 years ago?* BBLS Collection.

Below: *Once the remains of the ragged casing had been cleared away and scrapped 92 Squadron really did begin to look far more presentable. Indeed in this early view at Wansford she appears most business like. We do not have a date for this particular photograph, but you can gauge it from the fact that it appeared in Issue 8 of the* Steam Railway *magazine.* BBLS Collection.

An interesting comment was made in the December 1979 newsletter to the effect that, during 1980 we anticipated carrying out some shunting tests over one or two suspect sharp curves on the NVR with a full load of water and suitable ballast in lieu of coal. It was noted that this action could be risky and involve damage to the tender but would be preferable to taking the whole engine over them! This year also saw us investigating the procurement of a wheel drop to reduce the costs and difficulties of fitting and removing wheels by crane. Unfortunately the idea came to nothing.

In 1980 Dick Miles was officially voted in as Chairman at the AGM. On the locomotive front the engineers removed the wheels and cleaned the frames ready for crack testing on the frames, axles and boiler. The trailing axle evidently proving suspect caused one fright, but in the end it turned out to be a well-hidden keyway! A Mr. C. A. Eyre of Non-Destructable Testers from Castle Donnington carried out this work. New Tender Brake blocks were machined and its main frames were repainted. Work on the boiler progressed in the form of the removal of a small tube for inspection, it came away fairly easily. On the basis that it wasn't totally rotten, the indication was that the boiler was in a pretty clean condition. Peering through the small hole other tubes appeared to be in the same condition and the good news was that the usual boiler sludge was below the lowest tubes, thus these were not suffering unduly either.

The boiler was re-painted again, and the main frames were prepared for grit blasting. The other good news at the time was Derek Butler's work on the procurement of safety valves, which passed their pressure test and were subsequently certificated. The valves are the Ross-Pop' type with modifications by Derek along similar lines to valves on South African locomotives. The valves, which were made at Swindon, were the largest the works had made since those built for No.92220 *Evening Star*.

The tender was re-wheeled following the return of the wheels from Swindon. This simple statement belies the fact that after an initial trial lowering it was discovered that the bearings we had were not the ones we thought them to be and differed considerably from the drawings! Two weeks machining brought the offending items into line; literally! Derek Butler then discovered that there were actually no less than five different patterns of bearing! Many of the missing parts had now been found or manufactured and the tender was now back in a more or less complete condition again.

Work on the caravan was coming to a conclusion, as it was now fully fitted out with sleeping accommodation for six, a fully equipped kitchen with fitted cupboards, a Sadia hot water heater and a fridge. It was also professionally wired with twin sockets throughout. The bathroom boasted a brand new Sadia water heater, a washbasin, and a new Dolphin shower unit. This simply turned up at my house one morning after a couple of 'phone calls!

The lounge had another three-piece suite, coffee table and a TV. The van was fitted with new carpeting throughout, it also had new curtains and all the walls were papered. The running gear, chassis and roof had all been repaired. Meanwhile the outside was repainted in - yes you've guessed it, Malachite Green! It was plumbed ready to connect to the mains on arrival at Wansford. The only item not fitted was a toilet as the sewage disposal at Wansford could not be guaranteed at that stage. The total materials cost to the society for the caravan refurbishment came to £139.14 thanks to enormous sponsorship from numerous companies. The estimated savings because of this were £1,375.00!

Do you remember the Gales of January 1981? I do (for even though I had moved to Shrewsbury by this time due to work commitments) I had a 'phone call from Dick in the midst of them! He had the difficult task of advising me that the gales that had caused havoc in the South East, had removed half the caravan roof! The rain finished it off. Fortunately we had taken the precaution of insuring it for £1,500, which we claimed.

When we were paid out for the caravan I gained some solace from the fact that the Society was well in the black over the situation, but many hours had been 'wasted'. Hours that could have been raising funds or spent working on the engine, and of course we still didn't have any accommodation. The idea was left alone and we have managed without a roof over our heads at Wansford ever since. In 1981 Wing Commander J. E. Rooum AFC RAF was asked to become our new President and it was an invitation that he duly accepted.

On the Engineering side the welcome involvement of Roy Green and his son sped work along at a previously unheard of rate. A cross-head splitter was manufactured, and both cross-heads were separated from their piston rods, broken studs were removed from the saddle, and the threads cleaned out in readiness for re-fitting to the boiler. The valves were stripped out only to find that one head had shattered completely, obviously in BR days. One wonders, was this why she was withdrawn? Fortunately there was no damage to the bore. The piston rods were removed for re-grinding as pitting had occurred where the rods had lain in the glands during the locomotive's stay at Barry. Derek Butler in his inimitable way applied his mind to the lack of spring loaded buffers for the tender. He eventually tracked down a pair 'surplus to requirements' in Swindon works. Mind you we did have to pay for them - £5.00 each + VAT.

By 1982 the Nene Valley Railway had acquired a 40-ton Ransome's steam breakdown crane which would be very useful to us in due course, particularly for re-wheeling the engine, once the driving wheels had returned from Swindon. Harry Frith (ex-Eastleigh erecting shop) was approached to carry out special machining work on the crank pin in the middle of the driving axle. The main axle boxes were re-metalled and ready for machining.

Above: *In earlier days when lifting facilities were not readily available, other well tried past methods had to be revived. This impressive view was taken prior to additional packing being applied to roll out the next pair of wheels. One can imagine the amount of time taken by this method compared to utilising a crane.* Alan Whenman

Below: *Here we see 23 tons of distinctive Bulleid wheels loaded up on Charlie Roads lovely old Foden articulated lorry with its swan neck low loader. This was a scene that was repeated several times for the numerous journeys to and from British Rail Works at Swindon.* BBLS Collection.

Above: *With the boiler out of the way work on the frames could progress as this view shows. The wheels having been restored and returned from BR's works in Swindon, allows for serious work to begin as the engine is put back together. Now inside the shed at Wansford the extensive cleaning and painting programme for the frames had almost been completed. On 13th October 1984, the refurbished springs await fitting to give us a 'rolling chassis'. Jim Wade.*

Left: *By 6th January1986 the tender had been put back together, well not quite, but at least in a condition that would suffice for the time being. Notice that the raves have been re-fitted as originally built. (Or so we thought!).*

One axle box continued to give severe problems, Roy Green did some sterling work on this to eventually overcome the problem. The drag-box, which had been sent to Wolverton Works for restoration was due back and would then be re-instated. The large number of fixing bolts required for this would be turned on machines that were now installed at the farm. The right hand crosshead was re-fitted following its return from Swindon, where the white metal surfaces had been machined flat and parallel. Unfortunately someone there had not drilled the oil-ways correctly. Swindon did not want to remove any further material to correct the mistake and advised that it would run perfectly well as it was.

Considerable debate between Alan Whenman and Roy Green led to the decision that, although it was not up to the standard of the left-hand unit, it was in better condition than those currently fitted to *Britannia* and *Blackmore Vale* and it could be fitted. The Boiler Insurance Inspector paid us a visit and pronounced the firebox to be in good condition. Although all the small tubes would still have to come out for visual examination of the lower side of the flue tubes which would then be ultra-sonic tested. The smokebox door was removed for zinc coating to be applied (to preserve it). Four coil springs (2 leading and 2 trailing) were ordered to replace broken ones. The specification of these had to be altered to rectangular, as the original section was now unobtainable. The cost of these was £372, whilst the trailing truck tyres were also re-profiled.

An idea to obtain a loan from the bank to complete the restoration was aired at the November committee meeting (now at the farm by the way) a figure of £26,000 was anticipated to be required. Over 10 years this was felt (with interest added), to impose an intolerable burden on the Sales Officers, Ian Bowskill and John Cowdrey (Kent Area). Another item organised this year was the Piston Valve Appeal with the target of £400 and a good start was made on this. Roy Green took on the task of repairing and renovating all the studding on the main frames. The valve chests were completed then boarded up to prevent ingress of foreign bodies whilst other work proceeded.

In 1983 we fully re-painted the mainframes, whilst the pistons, piston valves and cross heads were all overhauled and re-fitted. The mechanical lubricators and what was left of the copper tubing were removed. New tyres were fitted to the trailing truck, and repairs to the drag-box continued. Tender brake gear was fitted and tested, raves were built up to the original height (some years later they were checked and still found to be 6" short so had additional metal added).

The vacuum pipes, vacuum system and water feed pipes were fitted, and the buffers and rear coupling added. Around this time a debate ensued as to fitting the locomotive with air brakes, partly as the NVR had air-braked continental stock, but as several BR MKIs were subsequently purchased this reduced the need!

Yet it was strange that some 15 years later we would again be discussing the topic, when Railtrack had announced the phasing out of MKIs on the main line. Air-braking is therefore essential for any preserved locomotive going 'main-line', and No.35028 *Clan Line* became the first so modified!

The wheels having arrived back from Swindon were found to have cracks in the tyres, and this necessitated their return for remedial work. The valve gear was fitted to the oil bath, and a few cracks were found in this, but these were soon welded and the bath made good. The rocker shaft to the oil pumps was renovated with new steel bushes, and the mechanical lubricators were finished off. The boiler tubes were removed and the boiler moved over to the locomotive yard from the station bay where it had been inaccessible to the public. The first guesstimate of a return to steam was made for the end of 1983! Then a notice appeared in the autumn newsletter:

FOR SALE

PARTIALLY RESTORED BATTLE OF BRITAIN CLASS LOCOMOTIVE 34081, *92 SQUADRON*. PROSPECTIVE PURCHASERS ARE ADVISED THAT THEY WILL BE REQUIRED TO GIVE AN UNDERTAKING TO CONTINUE WITH THE SAME HIGH STANDARD OF RESTORATION AS BEING APPLIED BY THE PRESENT OWNERS. IT IS ESTIMATED THAT THE SUM OF £25,000 IS NEEDED AT TODAY'S PRICES TO COMPLETE THE PROJECT. FULL DETAILS OF RESTORATION WORK COMPLETED SO FAR MAY BE OBTAINED FROM THE PRESENT SOCIETY'S SECRETARY. REASON FOR SALE - CONTINUING LACK OF FINANCIAL SUPPORT FROM PRESENT MEMBERSHIP.

Well we didn't sell it and we did complete it - eventually! By 1984 the tender pipe-work was completed, and a drag-box was temporarily fitted for drilling purposes. The boiler was cleaned out and two barrow-loads of scale were disposed of! Meanwhile the Ramsey Silent Chain Company of Bolton machined the bronze main driving sprocket and supply chain.

Through our contacts at *Port Line* we gained a good deal on two brand new Davies & Metcalf injectors, as this company were prepared to make a minimum batch of ten if orders for them could be found. As there were other Bulleid Pacifics in a similar position the deal was agreed. This kept the costs down to '2 for the price of 1'! This type of deal was something we were to do again with the clack valves.

Meanwhile the driving wheels and frames were top-coated. Harry Frith again came to our assistance with an excellent re-furbishing tool for the big end journal. Roy Green machined the oil pots and lids. During the year our friends in the *Port Line* group approached us for help, as they were searching for rods for their loco. It transpired that the centre rod we had was in fact off *Port Line*! In turn we needed to modify it to suit our needs. Obviously now the situation could be changed and indeed was. So our contemporaries requested us to get the costings to produce a new rod and offered us a similar price in order that they could regain their original.

Above: A good crowd could usually be guaranteed when a lift of this type occurs. Here the locomotive is raised high by the NVR 40 ton steam crane to enable the driving wheels to be rolled back in after return from Swindon in 1982. Roy Tanner.

Below: A number of the items that had to be procured following their disappearance from Barry. Brass valves including a safety valve, Bulleid gauges with their unique black backgrounds and the now highly polished regulator arm. Sadly many of these items were again to be stolen in 1997. BBLS Collection.

In turn this almost paid for the three new rod forgings required for our engine. By 1905 the refurbished cab had been returned to Wansford, whilst the cab support frames were grit-blasted. The cab itself was then test-fitted with the aid of the coaling crane only to be removed again some time later in order to permit the re-fitting of the boiler. The trailing truck was also re-assembled, and the front bogie axle boxes repaired but the main effort was the re-fitting of the boiler, which was done using the large 40-ton steam crane. The front buffer beam was straightened out. (Initially we had thought that this had been caused as a result of bad shunting at Barry, but on investigation it became obvious the damage was much older and was probably an earlier BR incident, can anyone enlighten us on this matter?) At this point it was envisaged that 1986 would be the year for steaming! The reverser was reunited with the frames, having been worked on, restored and modified by Tom Morgan, a considerable technical task. New injectors purchased at £1,700 + VAT, and the cab frames were welded and drilled by Dave Baker using a ratchet drill with 7/8" bit. The trailing truck was fitted, whilst the leading bogie was worked on and white metalled. Bogie frames and axles were painted black on the outside, with the insides becoming bright red. The main frames were then drawn out of the shed and turned in readiness for the fitting of the bogie, for which the main frame was lifted on screw jacks. The front apron was repanelled and No.34081 was added as originally done. The boiler was given another coat of red-primer, and the cab was made ready for its final fitting and work on the ash-pan started.

Early in 1986 we tried to fit the Ash-pan, but we were quite unsuccessful. However we did get the see-through smoke-box re-plated and we also bolted down the boiler using 16 (1 1/8") bolts plus 50 further bolts to secure the diaphragm plate. Sand boxes and mounting pins were finished off, painted and fitted. Following several requests, the engineering team reluctantly re-fitted the smoke-box door to at least make '92' look a little better.

The issue of the air-smoothed casing was now being considered and as would be expected after being ravaged for 11 years by the sea air, all 1,780 square feet of it needed replacing. A fund was therefore set up to sell it at £1 per sq.ft. We also re-upholstered and fitted the driver's and fireman's seats and fitted the vacuum ejector but not without considerable difficulties. The boiler was prepared for a BR inspection by removing the large tubes and superheater header. During this work on the front end the smoke box door handles were taken off and dropped into the chrome-plating tank - and then forgotten about! We now have the best chrome plated door handles in the country!

The main steam pipe to cab manifold was hauled up with strenuous efforts by all concerned and fitted to the base of the main valve. The body of the valve was cleaned and prepared for fitting. Clack valves were purchased for £300 each, but (as mentioned earlier) this was only achieved at this price by buying a quantity of five.

This meant that one each was obtained for *Manston, Port Line, P. & O., Canadian Pacific* and ourselves. Singly we were quoted £500 each! The other major expenditure in 1986 was £750 for an approximate length of 1,424ft of copper piping for the lubrication system. Chains for the valve gears were being sought with some difficulty. The original pitch chain based on the 'Morse Rocker Link' was no longer available due to the British standard modifying to an American pitch. Much midnight oil was burnt solving this situation until a new set of designs was approved and produced by the Ramsey Silent Chain Co. Ltd.

We also enjoyed a Royal Visit during 1986 when HRH Prince Edward inspected our project during his visit to the Nene Valley Railway on the occasion of the opening of their Peterborough extension. The prince spent some time looking at our engine and photographs of the restoration project, but unfortunately we didn't manage to enroll our honoured visitor as a new member.

Alan Whenman, our Chief Engineer did a brief assessment of the future major work to be done at this stage, it makes interesting reading: -
1. Machining of all castings acquired for boiler backhead and all controls.
2. Machining of all rods and associated bearings
3. Refitting of boiler tubes.
4. Fabrication and fitting of casing.
The cost of this little lot - a mere £20,000 - was still to find.

On the engineering front we finally fitted the ash-pans in 1987, and along with this we managed the grate supports, damper doors and operating gear, and also the hopper doors.

The air-smoothed casing appeal was progressing reasonably well, but our Chief Engineer then became a full time employee on the NVR. After an initial missed heartbeat or two we soon realised it was actually a useful move for rather than losing him, as might have happened, he was still on the scene. Thereafter Alan was to be around most of the time and could be called on for advice, in addition to which he still found time to continue working on '92' as well. The Nene Valley Railway finished the Peterborough extension, but were experiencing considerable difficulties in manning the entrance gate to Wansford, consequently a considerable amount of revenue was being lost. After some discussion we agreed with the NVR to man the gate for them, (at the risk of us losing personnel to man sales stands). In return we would receive one-third of the takings. The plan worked well and with the NVR receipts dramatically increased and very useful funds were also finding their way into our kitty.

Having now worked on the project for some 13 years (including time at Barry Island) we needed professional advice as to the engine's status. This came in the form of 'BR Special Preliminary Examination (Form K)' on 27th November 1987. Colin Woods the boiler Inspector was apparently very impressed by the standard of workmanship on the engine, which was a wonderful fillip for our engineers as we entered 1988.

Above: *The chassis now back into 'Pacific' status is about to receive the boiler on 24th November 1985, again by courtesy of the NVR steam crane. Roy Green, John Arthur, Dick Miles and Alan Whenman in attendance.* Jim Wade.

Below: *On the 30th June 1986, HRH Prince Edward (centre) came to open the NVR's 'Into-City' Peterborough extension. After this he was taken on a conducted tour around Wansford yard and shed. We suitably spruced up the engine and fitted the nameplates and crest for the first time in 20 years.* Peterborough Evening Telepgraph

In this year we applied to the Charity Commission for Charitable Status, which would have benefits with regard to Tax and VAT payments. We were allocated Registration No.299140 as an 'Educational Charity'. Our Engineers completed the oil-trays and straightened the tender raves, only eight years after they had been fitted! The ash-pan continued to keep us busy but the three-row cranks were fitted. Charlie Young who had for the time being finished playing around with the boiler, now turned his attention to the tender drawbar.

However his attempt to couple the tender to the locomotive didn't go too well at all and after further investigation he discovered that the drawbar appeared a couple of inches shorter than it should have been! A bit more head scratching and searching through the drawings revealed the problem. Tenders on re-built locomotives had drawbars 2 inches shorter than those fitted on the unrebuilt members of the class! That's the penalty of pinching someone else's tender! Fortunately it was not to be a major problem!

Left: *By 9th May 1987 progress is apparent, and No.34081 is looking like a locomotive again. The tender was more or less back together again, but the engine was to stay like this for a long time whilst lots of small jobs were completed in and around the boiler, chassis and cab. Even so, at least the locomotive was recognisable and reasonably presentable. Richard Perkins.*

Top Right: *On 16th March 1987 the cab (built by Apex Engineering of Letchworth) is now finally in place. In this picture we see that it is being fitted out with the back plate equipment, and the bearers are in situ awaiting the arrival and fitting of the new wooden floor panels.*

Bottom Right: *The manufacture of the casing was going to be a long job but by 5th March 1995 much of it had been fitted. In this view the rods were also on, and most of the copper piping had been added for the lubrication system. The cab once again had the luxury of windows. Ian Bowskill.*

Bottom Left: *In any project like this, you make contact with some remarkable people, and one of these was the late-Ivo Peters, the great railway photographer who had helped us with pictures of our engine in BR days. Following a request for missing back copies of our Newsletter, this letter (printed in facsimile) was received by our then Editor Richard Perkins:*

Dear Mr. Perkins,
*   I am not a member of the Battle of Britain Locomotive Society, but because a painting of your locomotive was based on one of my photographs, you very kindly sent me copies of '92 Squadron News'. One of the first copies I received was No. 39 (Winter 1984), which I see you want - so herewith my copy. Please keep this - it means far more to you than me. During the last war - being grade 4 medically - I was turned down by the RAF, but eventually got accepted by the Royal Observer Corps.*
I retired after 26 years service, but I am still an honourary member. Please excuse my awful scrawl. I am writing this lying down as I have cancer of the spine and now have to spend more of my time flat on my back. May I wish you complete success in the restoration of 92 Squadron. I enclose a small donation towards the work - very sorry I can't make it more, but being ill is rather expensive these days.
Kindest Regards,
***Ivo Peters**, Bath, Avon'*

CHAPTER TEN
ONLY TEN MORE YEARS TO GO!

As we progressed into 1989 we entered our last decade, although we did not know it at the time. The three throw crank valve was fitted with some considerable difficulty. This was a very precise job, as was the fitting of it to the chain gear drive. This ungainly lump had then to be located into the sump, and it fell to poor old John Haydon to undertake this onerous task but he persevered with it and achieved excellent results. Fitting of copper tubing for the lubrication system was underway and work progressed on the damper operating gear. Dave Barker completed the cab doors, front cowl and the rubbing strip where the fall plate on the tender makes contact with the locomotive. At the time we were fitting the front cowl, we had donated (quite coincidentally) a cast BR pattern smokebox number plate. With this the front end really began to come to life.

Above: *This is an interesting way to check if the boiler is full, wet feet - yes - dry feet - no! A cheery bunch carry on the good work in 1987. The tender was now coupled after considerable head scratching over its refusal to mate up. But this was our penalty for pinching a tender from a rebuilt loco.* Manni Masons Pictures.

The *Port Line* group came with a request to borrow our tender for some main line running with No.34072 as part of the Battle of Britain celebrations. Although this plan came to nought, it did spur us on to announce another completion date, this time 15th September 1990. The Ramsay Silent Chain company advised us that their work was progressing well, and that we could expect delivery of the coupling rod bearings and big end brasses by February 1990. Meanwhile work continued on the valve motion and the main Klinger valves. The super-heater header was fitted, whilst work was completed on the cylinder relief valves and the levers for the dampers and hopper doors. The wheels had to be removed once again, this time in order to rework the axle-box clearances.

Early in 1990 the sheeting for the casing began to arrive. Meanwhile the axle-boxes were built up with bronze welding and sent to Hull for white metalling and boring, after which they were duly returned to Wansford. Much welding was done on the tender, and Roy Holland painstakingly polished the coupling rods. The wooden cab window frames were made, but interestingly we found that the one on the left-hand side of the cab was 1/2" bigger in one direction than that on the other. The steaming date of 15th September 1990 came and went, and we went into the winter with numerous big jobs still to do!

At long last, on 25th April 1991, our engine was returned to the Pacific wheel arrangement, as all the wheels were fitted to make it a 4-6-2 once again. During the years since its rescue from Barry, the engine had masqueraded as an 0-6-2, an 0-6-0 and even a 4-0-2 for a while. Main springs, brake gear and boiler stud overhauls were all completed, and the steam generator had been worked on. Sanding pipes, sand trap and ejectors were all fitted in 1991, as were the oil pipes to the oil trays. The controversial Bulleid reverser, reworked by Tom Morgan, was fitted and tested but unfortunately did not live up to expectations. Time to correct what had been done would have detracted from the engine's completion programme, so we therefore decided to revert to the original design. The whistle shaft was fitted to the cab roof. On the running gear, the coupling rod bushes were re-metalled and bored by Plenty's of Newbury. The centre big end was white metalled ready for boring, with guidance from the owner of No.34071, Geoff Gore, who was employed by them.

By 1992 the gravitation lubrication system was finally completed, so that the drive chains to valve gear and oil pump were both installed, this in turn allowed the left hand coupling rods to be temporarily fitted. The coupling rod bushes were now white metalled. The large Klinger valve was fitted to the manifold, whilst the small Klinger valves were worked on. The Bulleid electric code lamps were ordered - all twelve of them!

The next year, 1993, saw a lot of work on the copper pipe runs along each side of the locomotive, whilst the steel framing for the casing came under Alan's scrutiny along with the damper door linkage.

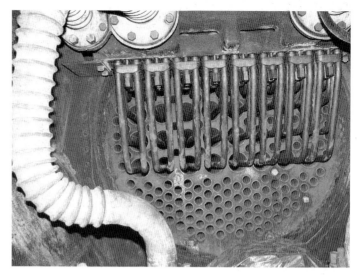

Above: *Here we have the first of two interesting pictures as the front tube plate shows off its newly fitted superheater elements and main steam pipes. These were convoluted to reduce fracturing from vibration. The white on the pipes is from ultra-sonic testing procedures. Phil Marsh.*

Below: *The boiler backhead showing Klinger Valves, gauge frames and copper pipe work in position. Phil Marsh.*

Also in 1993 a new drain cock was developed using Klinger packings to replace the taper, as this had the difficult task of keeping the steam tight, yet still had to be capable of rotation. Down on the farm, work was progressing on the smoke deflectors, Klinger Valves, displacement lubricator, fall plate and oil bath sumps. By 1994 a steaming date was still some distance away, but at least the engine looked as it should, following the extensive work that had been done on the casing that year. The cab windows were also completed, as was all the copper piping. It was decided that in order to make progress toward completion, John Haydon should compile a list of 'final' jobs, as the end was now really coming in sight.

Over any lengthy period like that involved with the restoration of *92 Squadron*, it is inevitable that there will be those who 'give up' along the way. More poignant, however, are those who pass away and 1995 was a bad year for the Society in this regard. The saddest time in our history was experienced in quick succession, following the death of Doris Miles, our Chairman's wife, who suddenly passed away on 11th December whilst her husband was out. It was an enormous shock to us all, but none so more than Dick himself. He and his wife were entirely devoted to each other, and Dick never got over the trauma, and he too passed away on 21st June 1996. This book is dedicated to their memory.

Following this dreadful situation the Society had to go on of course and we gradually picked ourselves up. Even so, our committee meetings were not the same any more without Dick's steadying influence in debates, and his down to earth, logical solutions to problems. Just as important was the loss of Doris's ever smiling face as she cheerfully put up with a rough load of rail nuts in her front room every month for years on end. We moved venue obviously and ended up at the 'Brewery Tap' Shefford, where John Haydon took over the Chair. By this time John had spent over 18 years on the project, and that has to be viewed as real dedication by anyone's standards

As 1996 progressed, steaming was once again being talked about, but seriously now. So many small jobs from the final list had now been tidied up, that by March work had been completed on the following items; pipe-work to the steam generator, vacuum ejector, steam brake valve, sander feed pipes, oil pressure pipes and cab gauges. The electrical lamps were all fitted and in the process of being wired up, the all-important regulator was fitted.

As these tasks were steadily being ticked off the list, the debate about main line running came up again. We all wanted this of course, but the reality of the situation begged caution. Upwards of £15,000 would be required on top of that already expended, plus the very topical problem of preventing lineside fires. This issue would have to be addressed by conversion to oil-firing, or the introduction of spark-arresting equipment, both of which were costly options.

At the time several preserved steam locomotives were undergoing the latter conversion, and trials were then in hand over Shap summit in Cumbria in order to ascertain the effectiveness of it all. Obviously the situation with the main line market was in a state of flux, especially with steam bans in operation during the summer months and additional expenses being levied by Railtrack for light engine movements. Add to this the fact that many rail tours were being cancelled (for a variety of reasons), and it soon put the economics of the additional expenditure into perspective. The final requirement of a 'support coach', with the associated costs, made us decide that we needed to get our feet wet on private lines first, although we also opted to 'keep tabs' on the situation over main line running as a future consideration.

As the year progressed the 'final list' got smaller and smaller, and even the big job of work on the casing was nearing completion. The largest outstanding task, and one of the most taxing then commenced with the re-tubing of the boiler. The preparation work for this job on the tubeplates was onerous, as we had to ensure that they were clean and free of old bits of tube and weld. Even so it was completed in time for the experts to come and load the new tubes into the boiler.

The new tubes were placed back into their rightful positions with the expert work of Chatham Steam Restoration Ltd., who then progressed with expanding the ends and welding them into position. On hydraulic test a few of them were found to weep a little but C.S.R. Ltd. soon popped back to sort out the offenders.

Having reverted to the original type of reverser, we found that this functioned satisfactorily when tested on air. Meanwhile a whistle was kindly donated, and once this was fully piped up it was checked mechanically and pronounced operational. The superheater header seats were refaced, the blower pipe was assembled and then fitted, after which the vacuum ejector exhaust pipe was finished. The steam generator under the cab was run up (on air), and the new lights all worked beautifully.

The impetus to complete was rapidly gaining ground, so many times we had announced that we were going to steam on such and such a date, but this time it was for real. More manpower had been placed at Alan's disposal and it was literally 'working'. Money was still desperately needed to accommodate this extra work load, as some of the labour had to be paid for. The freight charters were doing well, our beer trains were bringing an excellent return, and although the sales stands were spasmodic every little helped. A repayable loan of £6,000 was offered from two members on the basis that we would be in steam within a year. We were of course progressing nearer and nearer steaming and March 1997 was now being talked about as the (next) definite date! Towards the end of 1996 I moved back to Letchworth from exile at Carnforth, and was duly co-opted on to the committee after a 16-year break! It was like a time warp for many of the same faces were still there!

With our expectations hopelessly wrong again, another steaming deadline had passed, as the lack of manpower and financial resources still dogged our progress. We clearly recognised that more hours were again required on the engine. 'Egg on Face' was staring at us, we just couldn't delay any longer, no matter how justifiable the reason, yet at the same time we simply couldn't afford 'commercial' rates to employ more men. The solution came from John Haydon who suggested to the Committee that he should step down from being Chairman and Trustee. Then, in addition to his two days a week voluntary work, he could carry out a further three days as a paid employee (at a very reasonable rate it should be noted). This move was executed and with the Chair now vacant, I was voted in to hold that situation until the next AGM. With John's efforts supporting Alan the work continued apace, with the result that Sam Foster the BR boiler inspector was invited in on Friday 21st November 1997 to inspect a Hydraulic test.

Above: Prior to the Re-naming Ceremony on 12th September 1998, all hands prepare '92' for her big day.

On pronouncing the engine sound he then advised us that we could go for a steam test as soon as we liked. On Monday 1st December the first match was lit and a fire put in the firebox, so we honoured the conditions of our £6,000 loan! Furthermore, this was the first time No.34081 had had a fire in her since 1965. As this was 33 years ago, caution suggested a very slow warming up process in order to protect the boiler and its components. On Thursday 4th December the process was completed and the safety valves began to lift at full pressure, (250lbs/sq/in.) The steam test date was set for 12th December, when Sam Foster passed her with flying colours and the comment 'get the thing going as quickly as you can!' Getting her going involved the fitting of the super-heater elements, the regulators, lagging the boiler, refitting the boiler cladding sheets, and all the steam pipes and valves.

Top Left: *Do we take our hobby too seriously? Perhaps John Arthur had not realised this was not the way to check if the boiler was really empty. Whatever the case, he also has considerable faith in the locking mechanism of the Ajax fire hole doors!* Manni Mansons Pictures.

Bottom Left: *After 20 years of toil and graft, Alan Whenman has the proud (and very apprehensive) moment of lighting the fire for the very first time. It was a worrying time for all concerned, but before long No.34081's firebox was finally alight! The date was Monday 1st December 1997 and it was the first fire that had been in her for 33 years. Before long she would be steaming, a moment that many thought would never come after her purchase from Barry scrap yard. The warming-up process could not be hurried, but after all these years no-one was inclined to rush things and create a major set-back.* Ian Bowskill.

As the new year started, the date of the official launch was set for 24th March 1998 when we asked the record producer and locomotive owner Peter Waterman to officiate, but yet again a cruel blow was to take place and throw up yet another set-back! Having never suffered any theft from the locomotive or our adjacent stores vans, it came as a crippling blow to find that we had been robbed just after Christmas. The NVR night watchman was away at the time and the thieves did well for themselves, accumulating mainly non-ferrous material items with a face value of £25,000 from Wansford Yard. Fortunately(?) only £7,000 was stolen from ourselves, but they also took a whole variety of valves and other valuable brass and copper parts.

The sad thing about this was that the thieves could have only received a fraction of the true value in scrap prices. The net result of this was inevitable delays and rapid scouring of stores and other railways to find out if any willing donors could effect an immediate replacement whilst we had new items produced. Fortunately both the NVR and ourselves were fully insured for the losses. Undaunted we pressed on with fitting the replacements and completing the many other small jobs that were then outstanding, after which we planned to run-in the engine with the casing still awaiting completion. On 9th March 1998, she moved under her own steam for the very first time in 33 years. The only slight worry was a warm driving axle box, but this was corrected by adjusting the springs. Several days of gentle running up and down the yard at Wansford followed in order to prepare No.34081 for her first trip down the line to Peterborough.

I was in continuous discussion with the NVR Chairman, Martin Sixsmith, during this period in an attempt to iron out an agreement. The draft agreement was altered several times, but both sides eventually accepted it and all parties were happy. The locomotive was now ready to roll, fully repainted in her Malachite Green livery, the Sunshine lettering and the yellow stripes. Complete with the brass lamps and burnished metal, she looked absolutely wonderful. The signing took place at 11.30am on 23rd May 1998 and this was followed by No.34081 drawing her first passenger train out of Wansford Station just 15 minutes later. She then proved to be the star of the show at the NVR Spring Gala on the 6/7th June.

We had finally done it, she performed beautifully the whole of the weekend and continued working on the NVR throughout the summer. In addition to service trains, she was employed on driver training, photographic charters, and even a wedding special. The postponed official launch was then set for 12th September 1998, Battle of Britain weekend, this was to be held at the NVR and not the Bluebell (who had requested her, however her debut there was not until October). Negotiations complete, I began to feel under the weather and was subsequently advised by my doctor to take things a little easier. At this I decided to stand down from the chairmanship, and Ian Bowskill was subsequently voted in to take over the reigns.

Top Right: *The 23rd May 1998 was a significant day for the Society, the Agreement had at last been sorted. Signing was completed at 11.30 am, 34081 left Wansford on her first Revenue earning train at 11.45! Here the author, (then Chairman), shakes the hand of Martin Sixsmith, the NVR Chairman in the office at Wansford.*

Right: *Spring is traditionally a time of new birth, and for 34081 this came in 1998. Running light engine trials, 92 Squadron reverses through Wansford Station on 28th March 1998, showing the final form of the tender complete with the Bulleid electric lamps and further raised side raves. These were still found to be too short after the initial modifications way back in 1985. However the engine was back in steam, and although there was still much to do, this was a wonderful day for members of the Battle Of Britain Locomotive Society, and all those who helped us along the way.*

Although originally dubbed launch day, by the time 12th September came along, 34081 had been running on the Nene Valley for three months (virtually the whole of the summer). Accordingly the event was changed to a re-naming ceremony. Much behind the scenes work went on to ensure that all the details were covered, predominantly a lot of work fell on our engineering team who spent several days cleaning, polishing and burnishing the engine prior to the event.

Several RAF 92 Squadron veterans were contacted, but we had to make arrangements for these men as they were all well advanced in their years. Relatives of O.V.S. Bulleid were invited, but his son H.A.V. Bulleid had to decline on the grounds of the distance involved, although his daughter-in-law Jane Bulleid and two grandsons (David and Oliver) were happy to attend. Pete Waterman agreed to officiate again. Various RAF personnel arrived, including Spitfire pilots Dennis King, Barrie Cross and Len Stilwell, along with armourer Marson Peet.

Pilots from the later jet era were represented by ACM Sir Robert Freer, John Butler, Maurice Butler, Brian Lamb, Don Cartwright, 'Spike' Jones and Kevin Hutchinson.

At precisely 10am, 34081 crept quietly into the platform at Wansford Station, just as it started raining. Speeches were made by Martin Sixsmith, Ian Bowskill and Pete Waterman who unveiled the nameplate and crest. An RAF padre, the Reverend Group Captain Ron Hesketh, then re-dedicated the locomotive. At this point, and unknown to our guests, a Mk Vb Spitfire roared over and gave a superb flying display.

I do not think that anyone will be ashamed to admit there were tears in the eyes of many on the platform that day, as everyone was remembering! Some were remembering the dreadful days when they were up there in earnest during World War II. Some were remembering their family history and their now famous ancestor. Some were remembering the years of toil to achieve what was being experienced that day.

Left: *On the 23rd May 1998, No.34081 is caught powering through Castor towards Peterborough with her first return passenger working. The picture shows how important it was to get the correct height on the tender raves, otherwise the top yellow line would have been quite a problem.*

Top Right: *On 12th September 1998 came the wonderful day of the re-naming ceremony. No.34081 was wonderfully turned out, her casing gleaming and all the finishing touches added to the locomotive. You could almost have imagined that it had been prepared for working one of the crack Southern Region express services in their heyday. Once she pulled into Wansford Station, Ian Bowskill (BBLS Chairman) and Martin Sixsmith (NVR Chairman) gave speeches at the re-naming ceremony, after which Reverend Group Captain Ron Hesketh re-dedicate the locomotive. The group are seen here with Pete Waterman unveiling the nameplate.*

Bottom Right: *The highlight of the re-naming ceremony on 12th September was undoubtedly the fly past of a Spitfire, despite the rain and overcast sky a superb display was achieved bringing tears to the eyes of many assembled there. Although it forms just a small blob in the sky in our picture, the fly-past was a particularly important event. We were particularly fortunate to have a MkVb Spitfire for this event, as 92 Squadron were the first unit to be equipped with this version of the aeroplane in 1941. The pilot for the fly-past was Charlie Brown, and the Spitfire was based at Earls Colne in Essex.*

Some were simply remembering the people who didn't survive to see the day! So, with a wipe of the eyes our guests then boarded the train, and at 10.45am *92 Squadron* pulled out of Wansford station for Peterborough and her next fifty years.

What these fifty years will entail we can not yet say, but amongst the early activities (apart from keeping the engine in good running order) is Ian Bowskill's planning for the eventual movement of No.34081 to other railways and of course future Main Line activity. Both will be necessary in order to provide income to not only maintain the engine but to ensure as much as possible is in the kitty for the year 2008 when it starts all over again.

Hopefully at this point when the boiler certificate runs out she will be in far better condition than when those aspiring lads, Tony, Andy and Dennis first set eyes on the rusting hulk that was **No.34081 *92 Squadron*** all those years ago. I hope they and all the others will look back on the years of hardship and say, it was a job well done! I am sure that Oliver Bulleid would look upon them all as railwaymen after his own heart!

ACKNOWLEDGMENTS

In conclusion I must thank the many friends and colleagues who have contributed to this edition, in particular;
H.A.V. Bulleid,

and
Philip Atkins,
The Bulleid Society,
Professor Alan Earnshaw,
Mark Eaton,
John Fry,
The National Railway Museum,
The Nene Valley Railway,
Jim Oatway,
Barney Trevivian,
Ron White (Colour Rail),

and (of course) the numerous BBLS members who have contributed.

Left: *Shortly after her commissioning and still in the as-built condition , No.34081 (un-named) powers a van train through Wansford under the new signal gantry on 21st June 1998.* Ian Bowskill